Contemporary Concepts in Physics
Volume 2

L.B. Okun

Particle Physics
The Quest for the Substance of Substance

harwood

Particle Physics
The Quest for the Substance of Substance

Contemporary Concepts in Physics

A series edited by
Henry Primakoff
(1914–1983)

Associate Editors:
Eli Burstein
University of Pennsylvania
Willis Lamb
University of Arizona
Leon Lederman
Fermi National
Accelerator Laboratory
Mal Ruderman
Columbia University

Volume 1
Particle Physics and Introduction to Field Theory
T.D. Lee

Volume 2
Particle Physics
The Quest for the Substance of Substance
L.B. Okun

Additional volumes
in preparation

ISSN: 0272-2488

This book is part of a series. The publisher will accept continuation orders which may be cancelled at any time and which provide for automatic billing and shipping of each title in the series upon publication. Please write for details.

Particle Physics
The Quest for the Substance of Substance

L. B. Okun
Institute of Theoretical and Experimental Physics
Moscow, USSR

Translated from the Russian by
V. I. Kisin

h● harwood academic publishers
ap chur · london · paris · new york

© 1985 by OPA (Amsterdam) B.V.
All rights reserved.

Published under license by OPA Ltd. for Harwood Academic Publishers GmbH.

Harwood Academic Publishers

Poststrasse 22
7000 Chur
Switzerland

P.O. Box 197
London WC2 4DL
England

58, rue Lhomond
75005 Paris
France

P.O. Box 786
Cooper Station
New York, New York 10276
United States of America

Library of Congress Cataloging in Publication Data

Okun', L. B. (Lev Borisovich)
  Particle physics — the quest for the substance of substance.

(Contemporary concepts in physics, ISSN 0272-2488; v. 2)
Bibliography: p.
Includes index.
  1. Particles (Nuclear physics) I. Title. II. Series.
QC793.2.038 1984   539.7′2   84-10735

ISBN 3-7186-0228-8 (hardcover), 3-7186-0229-6 (softcover). ISSN 0272-2488. No part of this book may be reproduced or utilized in any form or by any means, electronic or mechanical, including photocopying and recording, or by any information storage or retrieval system, without prior permission in writing from the publishers. Printed in Great Britain by Bell and Bain Ltd., Glasgow.

CONTENTS

HENRY PRIMAKOFF: A REMEMBRANCE
Robert K. Mann   vii

PREFACE TO THE SERIES   ix

PREFACE   xi

CHAPTER I. PARTICLES AND PRINCIPLES   1

Experiment and theory, p. 1; Two trends, p. 2; Symmetries, p. 3; Relativity theory, p. 4; Action and Lagrangian, p. 5; Quantum mechanics, p. 6; Spin. Fermions and bosons, p. 8; Elementary particles, p. 9; Fundamental interactions. Hadrons and leptons, p. 10

CHAPTER II. GRAVITATION AND ELECTRODYNAMICS   11

Gravitation, p. 11; Quantum electrodynamics (QED), p. 12; The language of Feynman diagrams, p. 14; Polarization of the vacuum, p. 17

CHAPTER III. THE STRONG INTERACTION   19

Hadrons and quarks, p. 19; Isotopic spin. SU(2) group, p. 22; Strange particles, p. 24; SU(3) symmetry, p. 25; The charmed quark, p. 30; The b quark and others, p. 31; Flavors and generations, p. 32; Color and gluons, p. 34; Quantum chromodynamics (QCD), p. 36; Asymptotic freedom and confinement, p. 37; Chiral symmetry, p. 40; QCD on the march, p. 41

CHAPTER IV. THE WEAK INTERACTION — 43

Weak decays, p. 43; Weak reactions, p. 45; Components of the charged current, p. 46; Mirror asymmetry, p. 48; V − A current, p. 49; C, P, T symmetries, p. 51; Neutral currents, p. 53; Neutrino masses and oscillations. Double β-decay, p. 54; On the reliability of experimental data, p. 61

CHAPTER V. THE ELECTROWEAK THEORY — 63

Distinctive features of the weak interaction, p. 63; SU(2) × U(1) symmetry, p. 64; The photon and the Z boson, p. 65; Coupling of charged currents, p. 66; Coupling of neutral currents, p. 68; The search for the W and Z bosons, p. 68; Symmetry breaking, p. 70; Higgs bosons, p. 74; Models, models . . . , p. 75; Scalars: problem no. 1, p. 76; On the development of the theory, p. 77

CHAPTER VI. PROSPECTS FOR UNIFICATION — 81

Rendezvous of the running coupling constants, p. 81; Fermions in SU(5), p. 83; Gauge bosons in SU(5), p. 84; Proton decay, p. 85; Magnetic monopoles, p. 88; Models, models, models . . . , p. 91; Supersymmetry, p. 95; Unification models and the Big Bang, p. 99; On extrapolations and predictions, p. 109

APPENDIX 1. ON SYSTEMS OF PHYSICAL UNITS — 115

APPENDIX 2. GLOSSARY — 127

APPENDIX 3. BIBLIOGRAPHY — 179

On papers and preprints, p. 179; On conferences and schools, p. 180; Introductory comments to the list of reviews, p. 182; Sections and subsections of the list of reviews, p. 184; List of reviews, p. 186

SUBJECT INDEX — 219

Henry Primakoff: A Remembrance

Henry Primakoff, Donner Professor of Physics in the University of Pennsylvania and editor of this series, died at his home in July 1983, after a long illness during which he continued to teach and work at physics until the very end of his life.

Early in his career, Primakoff developed, with T. D. Holstein, the theory of spin waves which was based on a physical model and which employed theoretical techniques, both of which are now classics. Later, while at Washington University, he published a paper entitled "The Photoproduction of Neutral Mesons in Nuclear Electric Fields and the Mean Life of the Neutral Meson" which first described the process that became known as the Primakoff Effect and ultimately led to a precise measurement of the very short mean life of the neutral pi meson. That idea is still in use in explaining certain other particle production processes. Also while at Washington University, Primakoff and E. Feenberg were the first to suggest the possibility of a collapsed state of nuclei.

In the 1950's, Primakoff turned to the weak interaction as the focus of his research, elaborating the nuclear and particle phenomena that are manifestations of that interaction. He moved to the University of Pennsylvania in 1960, where he became a leading world authority on muon capture, double beta decay, and the interaction of neutrinos with nuclei. He continued to work in the fundamental symmetries of physics and the nature of their breaking until the time of his death.

Henry Primakoff was noted as a teacher as well as a researcher, giving freely of his encyclopedic knowledge of physics to students and colleagues alike. He had an abundance of patience and a joy in teaching that allowed him to place himself completely at the disposal of the individual with whom he engaged at the moment. This generous spirit earned him universal admiration and devotion.

HENRY PRIMAKOFF: A REMEMBRANCE

I had the pleasure and profit of many discussions with Henry Primakoff on a variety of topics. Occasionally, we wrote down some of this material and it was published. When one talked and worked with Henry, there was present the tension and excitement that is usual in an intellectual enterprise, but there was also a sense of tranquility, of deep enjoyment in the moment and the place and the exercise at hand. So there came to those of us who had the good fortune to work with Henry a sense of peace and a profound appreciation of the science of physics for its own sake. Henry Primakoff was indeed a rare individual to have bestowed such precious gifts on his friends and colleagues.

Alfred K. Mann

Preface to the Series

The series of volumes, *Concepts in Contemporary Physics,* is addressed to the professional physicist and to the serious graduate student of physics. The subjects to be covered will include those at the forefront of current research. It is anticipated that the various volumes in the series will be rigorous and complete in their treatment, supplying the intellectual tools necessary for the appreciation of the present status of the areas under consideration and providing the framework upon which future developments may be based.

Preface

Books popularizing physics naturally fall into two large categories conveniently referred to as "books for laymen" and "books for physicists." The former are dominated by the human, historical, aesthetic, and philosophical aspects of physics. The latter deal more with the physical substance of specific phenomena and laws. In the first category, authors avoid mathematical symbols: they might write "one millionth of one billionth of one centimeter." In the second, authors bravely operate with logarithms, exponents, derivatives, and integrals.

This is a book for physicists. Its potential readers are scientists, teachers, and students of physics and mathematics. It grew out of several talks at conferences in which I attempted to outline the general picture of elementary particle physics and to discuss its progress. In these talks, I addressed those who work at different sites of the construction of the Babel tower of high energy physics and do not always understand the language spoken at the adjacent sites. As a result, the book contains two "strata": a science-popularizing layer and a professional layer.

If you are familiar with special relativity but do not know quantum mechanics, you will understand roughly one-third of the book. If you have mastered the Schrödinger equation, you will understand one-half. If you can write the Dirac equation and know the meaning of the symbols in that equation, you will be comfortable with two-thirds. As for the remaining third, not everything is clear to the author himself.

This book consists of a Survey and three Appendices. The Survey contains six chapters: "Particles and Principles," "Gravitation and Electrodynamics," "The Strong Interaction," "The Weak Interaction," "The Electroweak Theory," and "Prospects for Unification."

Appendix 1, "On Systems of Physical Units," is mostly devoted to a discussion of the system $\hbar = c = 1$ and to a comparison with other systems, including the International System of Units (SI). One of the purposes I had in mind was to teach the reader to employ relativistic quantum units: the system in which $\hbar = c = 1$. This system clarifies the fundamentals of the phenomena and enables a physicist to carry out back-of-the-envelope dimensional estimates, many examples of which are scattered throughout the book.

Appendix 2, "Glossary," explains about one hundred terms. Originally, the Glossary was conceived as a collection of brief explanations related to the Survey and was meant for a more or less unprepared reader. However, time and again I was unable to resist a temptation to make a remark, or give a clarification, that could be of interest to a specialist as well.

It will be immediately noticed that many important terms were not assigned an individual entry in the Glossary. In most cases, they are explained in related entries or in the text of the Survey. The Subject Index will help the reader locate these explanations quickly. The Subject Index plays an important part in the book, creating bridges between sections of the Survey and Appendices 1 and 2, thus binding the book together.

Appendix 3, "Bibliography," is mainly a systematized list of papers on high energy physics and related topics published in leading popular-science and review journals after 1975.

I recommend that the reader start with a cursory reading of the book, not halting at difficult passages, only pencil-marking them in order to return to them at the second reading. It might be advisable to begin by looking through Appendices 1 and 2 and the Subject Index, and only then passing to the body of the Survey.

A few more words should be said to those readers who are only at the beginning of a serious study of physics. At the second reading of the book you should carefully look at the figures and, especially, at the formulas. Formulas do appreciate careful contemplation. When looking at a relation, compare the dimensions of different terms and compare their tensor indices. Ask yourself what is denoted by each letter or symbol in each specific expression. (I'll be grateful if you inform me about any misprints that your analysis may uncover.)

This book will not answer all the questions it may prompt you to ask. It will not be able to replace either a quantum mechanics textbook or, even less so, a textbook on quantum field theory.

The objective of this book will be met if it helps you to grasp the links between new theoretical concepts and experimental facts established during the past one or two decades and stimulates you to turn to more specialized books† and journals.

Mr. Martin B. Gordon, who visited Moscow two months before the Twentieth International Conference on High Energy Physics (Madison, USA, July 1980), suggested that an enlarged text of the summary talk that I was preparing at the time for that conference be included in a series of

---

†A few recent examples are: C. Itzykson and J.-B. Zuber, *Quantum Field Theory,* McGraw Hill, 1980; T. D. Lee, *Particle Physics and Introduction to Field Theory,* Harwood, 1981; and L. B. Okun, *Leptons and Quarks,* North Holland, 1982.

books he is publishing. I am grateful to him for this suggestion, which finally resulted in the present book. I did not suspect how much time the work would take when I accepted the proposal.

I am also grateful to I.Yu. Kobzarev, V.I. Kogan, A.B. Migdal, N.G. Semashko, K.A. Ter-Martirosyan, M.B. Voloshin, and many others, who read parts of the manuscript, for numerous comments (alas, only some of them are reflected in the final text).

My special thanks go to E.G. Gulyaeva and I.A. Terekhova for their assistance in preparing the manuscript for publication.

L.B. Okun

Chapter 1

PARTICLES AND PRINCIPLES

Experiment and theory
Two trends
Symmetries
Relativity theory
Action and Lagrangian
Quantum mechanics
Spin. Fermions and bosons
Elementary particles
Fundamental interactions. Hadrons and leptons

EXPERIMENT AND THEORY

The physics of elementary particles is a remarkable alloy of experiment and theory.

The properties of the minutest building blocks of matter were, and are, established in experiments whose complexity has no rivals in other fields of science. These unique experiments combine a truly industrial scale with a fantastic precision. The objects of study, that is, particles, are created in the laboratory with accelerators; their lives are so short that a blink lasts an eternity in comparison. One event of a rare decay of a particle must be searched for among billions of "dull" background events. All information on elementary particles is extracted by painstaking measurements.

But the hunt for, and the accumulation of, this information as such is not the goal; it is not the ultimate objective of the physics of elementary particles. The highest ambition of particle physics is to establish the most general physical laws of nature. The data provided by experiments must be transformed into theoretical conclusions. The quintessence of the theoretical analysis of many hundreds of experiments is the theoretical notions and mathematical formulas embodying them, which could be written on just a few sheets of paper. Ideally, this would be a single formula, the magic acorn containing the whole tree of physics, but we are still quite far from this ideal.

TWO TRENDS

Two opposite and, at first glance, even antagonistic trends stand out in the history of physics. On the one hand, there is the exponential growth of the number of phenomena under study, the ever greater specialization, and the ever finer branching of each direction of research. This process of branching, or differentiation, is especially well illustrated by the proliferation of new specialized journals and conferences.

On the other hand, the counterprocess, that of unification, or synthesis, or integration, proceeds just as intensely. Year by year, relationships between separate branches of physics, between phenomena which seemed to have nothing in common, become more discernible.

Newton's mechanics unified the motion of terrestrial and celestial bodies. Maxwell's electrodynamics unified electric, magnetic, and optic phenomena. Einstein's special relativity unified space and time. Quantum mechanics unified the concepts of particles and waves, and determinism and probability; as a result, it unified atomic physics with chemistry and the physics of condensed matter. Quantum field theory unified particles and forces. The progress in quantum field theory that we witness nowadays promises to unify various types of elementary particles and the fundamental interactions between them. Here I mean the so-called theories of grand unification and superunification.

Differentiation and unification may appear to be mutually exclusive, but only to a very superficial observer. Physics progresses by concrete accomplishments, and each step towards synthesis calls for more sophisticated and specialized means. This is true not only of experimental technique but also of the mathematical tools of the theory. In turn, each new stage of synthesis invariably generates diverse, far-reaching directions of research, not only in science *per se*, but also in technology, radically altering mankind's way of life. Suffice it to recall radio and nuclear technology. The former is a spin-off of the electrodynamic synthesis and the latter of the relativistic and quantum syntheses. It is very likely that the ideas of the grand unification and superunification theories will open no less breathtaking vistas.

Although, for physics as a whole, the process of differentiation and branching is indispensable, it nevertheless creates very serious difficulties for each individual scientist. As a result of the constant ramification of science at the front lines into ever newer directions, physicists of different specialties have difficulty understanding one another, even if they work at the same institute.

The more profound the understanding of the given subject, the more precise and rich the language describing it. Indeed, in science, this language is a tool of cognition. But the richer the language of a given branch of science, the more difficult it is for others to understand. A person choosing to be a polyglot in physics faces the risk of having neither energy nor momentum left for creative activity. Anyone doing research in science must fight at two fronts: to struggle both with nature and with his or her own ignorance. At the former front, new scientific knowledge emerges; at the latter, one tries to master what others have created. Both types of activity are inseparably interwoven.

This book was written to help those who want to form a general notion of the main ideas and trends in modern elementary particle physics. Its purpose is to offer help (even if not perfectly efficacious help) in overcoming the language barrier and thus to contribute to the unity of physics.

SYMMETRIES

The key notion of modern physics is that of symmetry. Symmetry is a tool by which it is possible to single out certain fundamental structures from among a kaleidoscope of physical phenomena and to reduce the diversity of the physical world to several dozen fundamental formulas.

A child notices and recognizes symmetry long before learning this word. A butterfly, a ball, alternation of day and night. . . . Many manifestations of a few distinct types of symmetry surround each of us throughout our lives. Physicists may be called hunters for symmetries: in a certain sense, they differ from other people only in that they search for ever better hidden and more fundamental types of symmetries in nature. In the final analysis, this is precisely where a physicist is led by his work, although he may not always be aware of this fact.

The notion of symmetry is inseparable from the notions of transformation and invariance: A ball is invariant under rotations, as the two wings of a butterfly are under mirror reflection . . .

RELATIVITY THEORY

The set of transformations forming the so-called Poincaré group is well known: it includes translations in space and time, spatial rotations, and motion at a constant velocity. The invariance of the laws of nature under these transformations is the essence of Einstein's special relativity theory. This invariance appears because space and time are homogeneous and the conventional three-dimensional Euclidean space and four-dimensional Euclidean space are isotropic (the latter is known to differ from the real physical pseudo-Euclidean Minkowski space in replacing real time t with imaginary time $i\tau$).

The invariance of the laws of nature under the Poincaré group transformation is manifested in the existence of a number of conservation laws: conservation of energy E, momentum **p**, angular momentum **M**, and the Lorentz momentum **N** (**M** = **r** × **p**, **N** = t**p** − rE for a pointlike particle with energy E and momentum **p** located at a space-time point with coordinates t, **r**). In an isolated island-like system of particles, the total values of E, **p**, **M**, and **N** are conserved no matter what the interactions within the system are.

The fundamental constant in the equations of special relativity is the limiting velocity of propagation of physical signals: the velocity of light, $c \approx 3 \times 10^{10}$ cm/s.

Under the coordinate transformations that form the Poincaré group, the quantities ct and **r** transform as components of a four-dimensional vector x_μ; similarly, E and **p**c transforms as p_μ, and **M** and **N** transform as components of an antisymmetric tensor $M_{\mu\nu}$ ($\mu,\nu = 0, 1, 2, 3$).

Some quantities remain unchanged under these transformations. These are invariants (scalars): the space-time interval, $s = x^2 = x_\mu x_\mu = (ct^2 - \mathbf{r}^2)$†, the square of the mass, $m^2 c^4 = \mathbf{p}^2 = p_\mu p_\mu = E^2 - \mathbf{p}^2 c^2$, and, finally, $M^2 = M_{\mu\nu} M_{\mu\nu}$.

No serious discussion of relativity is possible without considering physi-

---

†In these formulas, and hereafter, a pair of identical indices (called "dummy indices") denotes summation. In the case of four-dimensional indices, summation implies an additional minus sign in front of the products of spatial components. Therefore, a product of two four-dimensional vectors a_μ and b_μ is

$$ab = a_\mu b_\mu = a_0 b_0 - a_2 b_2 - a_3 b_3.$$

Clearly, the temporal and spatial terms differ in sign because Minkowski space is pseudo-Euclidean.

cal fields. The concept of the electromagnetic field as an independent object in physics was formed by Faraday, Maxwell, and others long before special relativity theory was developed. Only after this latter theory was born, however, did it become apparent that the introduction of the general concept of a physical field as a system with an infinite number of degrees of freedom, varying in space and time, was absolutely inevitable. Indeed, in the absence of instantaneous action-at-a-distance, any forces exerted by particles on other particles as a result of changes in the positions of the former particles can only be transferred as perturbations of the field propagating at a finite velocity from one point to another. Fields carry energy and momentum. Relativistic invariance demands that potentials of various fields be transformed in a specific manner under a four-dimensional rotation. Thus, the potential of an electromagnetic field, $A_\mu(x)$, is a four-dimensional vector. At the present time, physics operates with a large number of different fields. Some of them are vector fields like the electromagnetic field, that is, are described by potentials that are four-dimensional vectors. Scalar, tensor, and some other fields are also known.

ACTION AND LAGRANGIAN

Among all physical quantities there is one unique quantity that occupies the central position in physics. This quantity is the action S. For a free nonrelativistic particle with kinetic energy T_{kin}, the action from an instant of time t_1 to another instant t_2 is

$$S = \int_{t_1}^{t_2} T_{\text{kin}} \, dt.$$

In more complex physical systems,

$$S = \int_{t_1}^{t_2} L \, dt,$$

where L is the so-called Lagrange function for a nonrelativistic particle in a static potential,

$$L = T_{\text{kin}} - U,$$

where U is the potential energy.

For a field

$$S = \int \mathscr{L}(x) d^4x,$$

where $\mathscr{L}(x)$ is the so-called Lagrangian, $x = (ct, \mathbf{r})$ is the coordinate of the space-time point, $d^4x = (cdt, d\mathbf{r})$, and the integral is taken over the whole space–time.

The central position that action occupies in physics stems from the fundamental law of physics: the principle of least action, whose classical formulation states that in real processes observed in nature action is extremal (its variations vanish):

$$\delta S = 0.$$

The variational principle was introduced into physics by Fermat in 1662 ("nature follows the easiest and most accessible paths"); the notion of action (*actio formalis*) originates with Leibniz. The principle of least action was later elaborated by Maupertuis, Euler, Lagrange, Hamilton, and others. For a long time, however, this principle was regarded as a mere accessory to the Newtonian laws of motion. The universal role of the action in physics became clear only after the works of Helmholtz, Planck, and Noether.

The conservation of energy, momentum, and angular momentum follows from the invariance of action under the Poincaré group transformations. It will be shown later that other conservation laws stem from the invariance of action under other transformations. However, the true prominence of action lies not in conservation laws but in the fact that action contains all the dynamics of interactions between fields and particles. The least action principle applied to S and L yields equations of motion. Thus, it is often said that constructing a theory of elementary particles reduces to finding the fundamental Lagrangian describing the physical world and to solving the equations of motion it gives. Later, we shall discuss what fields and particles enter the fundamental Lagrangian and what the interactions between them are. We shall see that the guiding stars in the search for various terms of the fundamental Lagrangian are symmetries.

QUANTUM MECHANICS

Relativity theory is one of the two pillars supporting the edifice of modern physics. The second such pillar is quantum mechanics, created in the 1920s by Bohr, De Broglie, Heisenberg, Dirac, Schrödinger, Born, and others.

The fundamental constant in quantum mechanics is Planck's constant, or the quantum of action $\hbar = 1.05 \times 10^{-27}$ erg s.

According to quantum mechanics, the classical trajectory of a particle that moves from point A to point B is only one (usually the most probable) trajectory among a whole class of admissible trajectories. The particles themselves are no longer particles in the conventional classical sense; they possess wave properties that are the more pronounced the smaller the mass of the particle and the smaller the region of space in which the particle is forced to move by external forces. In quantum mechanics, particles and systems of particles are described by state vectors in an abstract space (Hilbert space). Dynamic quantities are put into correspondence with operators acting on Hilbert vectors.

Transformations of Hilbert vectors (multiplication by phase factors, linear transformations not related to any space-time transformations but taking place, so to speak, entirely within the Hilbert space) are called internal transformtions. These transformations correspond to "internal" symmetries, which are important in theories describing interactions between elementary particles.

Quantum-mechanical laws were first encountered in atomic physics. These laws are predominant in the physics of nuclei and elementary particles.

Elementary particle physics primarily deals with quantum relativistic processes in which the characteristic value of the action S is comparable to $\hbar$ and velocities v are comparable to the velocity of light c†. Frequently, the energy of particles exceeds their masses by many orders of magnitude. Under these conditions, the processes of particle creation play a paramount role.

The theoretical equipment for a description of this scope of phenomena is provided by quantum field theory. There are several equivalent formulations of quantum field theory. Most often, classical fields are put into correspondence with operators of the creation and annihilation of particles that are quanta of the corresponding fields (in the case of the electromagnetic field, these particles are photons). In this formulation, both the Lagrangian and the equations of motion derived from the least action principle are treated as operators.

According to another formulation suggested by Feynman, quantum-field dynamics is described by a functional integral over all field configurations,

---

†The theory of quantum relativistic processes is considerably simplified by using a system of units in which $\hbar = c = 1$; this natural system of units will be used throughout the book.

each configuration included with a weight $e^{iS/\hbar}$, where S is the action corresponding to the configuration. In the classical limit, the main contribution is made by configurations with extremal action. By using this formalism, Feynman elaborated the mathematical apparatus of Feynman diagrams which will be described in the next chapter.

SPIN. FERMIONS AND BOSONS

One of the most important laws of quantum mechanics is the quantization of angular momentum. The orbital angular momentum L can only be a multiple of $\hbar$; more precisely, the projection of L onto any of the coordinate axes assumes only the values $m\hbar$, where m is integral, and $-l \leq m \leq l$ with $l = 0, 1, 2 \ldots$

In addition to the orbital angular momentum, particles possess an intrinsic angular momentum called the spin. The spin of a particle is one of its inseparable and permanent properties. Particles with zero spin are called scalar, those with spin $\hbar/2$ spinor particles, those with spin $\hbar$ vector particles, those with spin $(3/2)\hbar$ spin-vector particles, and those with spin $2\hbar$ tensor particles. When the spin of a particle is mentioned, it is always understood that it is expressed in units of $\hbar$. For instance, the electron is said to be a spin 1/2 particle, while the photon is a spin 1 particle.

Depending on the value of the spin S†, all particles are divided into two large classes: particles with half-integral spin ($S = (n + \frac{1}{2})\hbar$, where n is an integer) are called fermions and particles with integral spin ($S = n\hbar$) are called bosons. A given quantum-mechanical state may contain an arbitrary number of bosons but only one fermion of a given type. It is this last law, the Pauli principle, that governs the filling of electron shells of atoms.

The existence of fermions signifies that the spin of a particle cannot be reduced to the orbital motion of its components. Spin is the most important property of matter that is not yet completely understood. A description of the spin by the mathematical tools of group theory served as a prototype for various theories of the "internal" symmetries, including the simplest of them, the theory of isotopic spin. The objective of the so-called supersymmetric approach is to construct schemes of a symmetry unifying fermions and bosons. This and other aspects of sypersymmetry will be discussed later.

---

†Sometimes the spin is denoted by the letter J.

Now it is time, at last, to define what is understood by the term "elementary particles."

ELEMENTARY PARTICLES

Elementary particles are usually defined as particles that cannot be decomposed into constituents. This definition excludes atoms and atomic nuclei but subsumes electrons, protons, and neutrons. Electrons form atomic shells: Protons and neutrons form atomic nuclei (protons and neutrons are both referred to as nucleons).

It will become clear further on that nucleons comply with the naive idea of elementarity to a much lesser degree than electrons. Nucleons have quite appreciable size (of the order of 10^{-13} cm) and a complex internal structure. It should be said that the term "elementary particles," like quite a few other terms in physics, should not be taken at its face value.

Another abundant and quite familiar elementary particle is the particle of light, the photon. Only slightly less abundant, but much more obscure to the public, are neutrinos. These electrically neutral particles are very difficult to observe because their interaction with electrons and nucleons is extremely weak; thus, they pass through tremendously thick layers of matter, without interacting.

Neutrinos ν, photons γ, electrons e, and protons p are stable particles; either they do not decay at all or decay but at a very low rate. (The experimentally established lower limit on the lifetime of the electron is approximately 10^{22} years and that of the proton 10^{31} years, much longer than the lifetime of the universe, 10^{10} years.) A free neutron n decays in approximately 10^3 s, but neutrons bound in nuclei are stable.

In addition to these stable particles, we know of several hundred unstable particles whose lifetimes lie in the range of 10^{-24}–10^{-6} s. Most of them live less than 10^{-20} s; they are called resonances. (In order to distinguish the other, "long-lived," particles from resonances, physicists often refer to the former as stable particles. Thus, well-known *Particle Data Tables* combine the truly stable and the long-lived quasistable particles into a single table under the heading "Stable Particles.")

A collision of two sufficiently energetic particles creates many new particles. Hundreds of particles are created in some of the collisions observed, yet these particles are not fragments of the colliding two, but full-fledged new-born particles. Nature "mints" particles under many diverse conditions but, regardless of the mode of minting, all particles of a given type are as alike as new dimes and do not wear out at all until the moment of death, that is, decay. It is impossible to break a "chip" off an elementary particle. A decay of an unstable elementary particle creates lighter particles, but these decay products are not components of the decaying one: they are created at the moment of decay.

FUNDAMENTAL INTERACTIONS. HADRONS AND LEPTONS

Elementary particles participate in a virtually unlimited number of processes but there are only four types of fundamental interactions underlying all the processes observed up to the present day: gravitational, electromagnetic, weak, and strong.

The gravitational interaction is universal: It involves all elementary particles. The source of a gravitational field is the energy-momentum tensor. The sources of electromagnetic field are electric charges. Neutral particles carrying no electric charge interact with the electromagnetic field only because of their complex structure (or because of quantum effects). In this sense, the electromagnetic interaction is not as universal as the gravitational interaction. To a certain extent, this is also true for the weak interaction. Only particles called hadrons participate in the strong interaction. Actually, the majority of elementary particles *are* hadrons. In addition to the proton and neutron, the hadron family includes numerous mesons and hyperons, both long-lived and resonances.

Only six fermions are known not to participate in strong interactions. These are the so-called leptons: the electron e, muon μ, tau-lepton τ, and their corresponding neutrinos (ν_e, ν_μ, ν_τ).

Theorists conjecture that interactions may not be limited to the gravitational, electromagnetic, weak, and strong; other types have been predicted, but have not yet been seen, despite numerous searches. Some of these hypothetical interactions will be discussed in Chapter VI; first, however, we shall discuss the interactions that are known quite well.

Chapter 2

GRAVITATION AND ELECTRODYNAMICS

Gravitation
Quantum electrodynamics (QED)
Language of Feynman diagrams
Polarization of the vacuum

GRAVITATION

The nonrelativistic theory of gravitational interaction, created by Newton three centuries ago, is the earliest physical theory in the modern meaning of this word. The universal gravitational action at a distance between two bodies with masses m_1 and m_2 is described in this theory by the potential

$$-G_N \frac{m_1 m_2}{r}$$

where G_N is Newton's gravitational constant: $G_N \approx 6.67 \times 10^{-8}$ cm^{3}g^{-1}s^{-2}.

The relativistic theory of gravitation, that is, general relativity theory, was developed by Einstein on the basis of the idea that the theory must be uniquely determined by the requirement of invariance under local transformations. In general relativity, this means the invariance of equations under arbitrary transformations of four-dimensional coordinates that are different at different world points. The form of the action in general relativity is established by using this principle of universal coordinate invariance.

General relativity has subsumed the Newtonian theory and predicted and quantitatively described a number of new effects: the deflection of light rays and radio waves by a gravitational field (that of the sun), the precession of Mercury's perihelion, gravitational waves, and black holes. The role of general relativity as the basis of modern cosmology is especially important in the Friedmann expansion of the universe and the theory of the primordial big bang.

Unfortunately, attempts to construct a quantum theory of gravitation have so far been unsuccessful. This is mostly due to two factors. The first is that gravitational interactions between individual elementary particles is very weak and thus lies beyond the capabilities of experimental study in the laboratory. Suffice it to say that Newton's potential was not even checked at distances below 1 cm. As a result of the weakness of the gravitational interaction, gravitational waves have not yet been observed, and the observation of individual quanta of gravitational field—that is, gravitons—seems to be a problem that the experimentalist will hardly be able to solve even in the next century.

The second stumbling block in the way of a quantum theory of gravitation is this: It must be the most complex of all known physical theories, because the complexities in a quantum relativistic theory increase sharply as the value of the spin of the particles described by the theory increases.

It can be shown that, since the graviton has spin 2, the gravitational interaction due to the exchange of gravitons increases with increasing energy and becomes strong at energies of the order of $m_P c^2$, where m_P is the Planck mass:

$$m_P = \sqrt{\hbar c / G_n} \approx 1.22 \times 10^{19} \text{ GeV } c^{-2}.$$

Attempts to calculate the forces due to the exchange of two or more gravitons lead to meaningless infinities (divergent integrals).

Summarizing, we can say that a quantum theory of the gravitational interaction has not been constructed because it is too weak (at available energies) and too strong (at energies of the order of $m_P c^2$).

As for the Planck mass, we shall see later that it may define the scale of the whole of fundamental physics.

QUANTUM ELECTRODYNAMICS (QED)

The electromagnetic interaction is the interaction of electric charges with an electromagnetic field; it has been much better studied than all the other fundamental forces in nature. This is not surprising. Indeed, the electromagnetic interaction is the basis of almost all the processes and phenomena around us: physical, chemical, and biological.

Quantum electrodynamics (QED) is the theory of the electromagnetic interactions of electrons and positrons; it is the most accurate of all physical theories. Here the electromagnetic interaction is found in its pure form. The unparalleled accuracy of calculations in quantum electrodynamics is based on the techniques of perturbation theory, using a small dimensionless parameter $\alpha = e^2/4\pi\hbar c \approx 1/137$, where e is the electric charge of the electron. The most advanced calculations are the computations of the magnetic moment of the electron: they take terms of the order of α, α^2, and α^3 into account. These computations are in excellent agreement with measurements. The experimental and theoretical values of the magnetic moment agree to nine decimal places.

Quantum electrodynamics provides excellent descriptions of the electromagnetic properties not only of electrons but also of two other charged leptons: μ and τ. In contrast to this, the electromagnetic properties of hadrons are determined, to a large extent, by strong interactions; their complex structure makes the calculations much harder. Electrons and muons are used as electromagnetic probes to study the internal structure of hadrons. Of special interest are the so-called deep inelastic electromagnetic processes at high energy and large momentum transfer; for instance, the annihilation of e^+ and e^- into hadrons observed in electron-positron colliders or the multiple production of hadrons in collisions of high-energy electrons or muons with nucleons.

From a purely theoretical standpoint, the importance of quantum electrodynamics can hardly be overestimated. It is the simplest and best studied chapter of quantum field theory. It was in the framework of quantum electrodynamics that many fundamental concepts and principles of that theory were discovered and formulated. It serves as a model for the construction of more complicated theories of strong and weak interactions, as well as grand unification theories.

The foundation of quantum electrodynamics was laid by Dirac at the end of the 1920s. The theory was cast in its present form at the end of the 1940s and beginning of the 1950s by Feynman, Schwinger, Tomonaga, Dyson, and others.

Quantum electrodynamics introduced the first antiparticle: the positron. It was in the framework of QED that physicists first realized that particles and forces are two manifestations of more complicated objects: quantum fields described by operators. Thus, the operator $A_\mu(x)$ creates or annihilates a quantum of the electromagnetic field at a point x, the operator $\psi(x)$ annihilates an electron or creates a positron, and the conjugate operator $\bar\psi(x)$ annihilates a positron or creates an electron. The Lagrangian of QED is a sum of the local products of these operators (the term "local product" means that the operators in a product refer to the same space-time point):

$$\mathcal{L}(x) = \bar{\psi}(x)\,[(i\partial_\mu + eA_\mu)\gamma_\mu - m]\,\psi(x) - \tfrac{1}{4}F_{\mu\nu}(x)F_{\mu\nu}(x),$$

where $\partial_\mu = \partial/\partial x_\mu$ is a partial derivative with respect to the coordinate x^μ, $F_{\mu\nu} = \partial_\mu A_\nu - \partial_\nu A_\mu$ is the operator of the electromagnetic field strength, e is the electric charge, m is the electron mass, and γ_μ are four Dirac matrices; a repeated subscript, as always, implies summation. The first and third terms in the Lagrangian describe the free motion of electrons and positrons, the last term describes the motion of photons, and the term $\bar{\psi}A\psi$ describes the interaction of photons with e^- and e^+.

If we introduce the so-called covariant (or "long," in students' lingo) derivative

$$D_\mu = \partial_\mu - ieA_\mu,$$

then the Lagrangian of QED takes the form

$$\mathcal{L} = \bar{\psi}[iD_\mu\gamma_\mu - m]\psi - \tfrac{1}{4}F_{\mu\nu}F_{\mu\nu}.$$

The "shorter" derivative ∂_μ and the 4-potential A_μ enter the Lagrangian only through D_μ and $F_{\mu\nu}$.

It is easily verified that the Lagrangian of QED is invariant with respect to a so-called gauge transformation,

$$\psi(x) \to e^{i\alpha(x)}\psi(x),\quad \bar{\psi}(x) \to \bar{\psi}(x)e^{-i\alpha(x)},\quad A_\mu(x) \to A_\mu(x) - \tfrac{1}{e}\partial_\mu\alpha(x).$$

The gauge symmetry of QED is responsible for the masslessness of the photon.

The gauge symmetry of QED is called Abelian because, in this case, two successive transformations commute (the result is independent of their order). In the cases of the strong and weak interactions (to be described below), we again deal with gauge transformations, but with non-Abelian ones: they do not commute.

THE LANGUAGE OF FEYNMAN DIAGRAMS

Calculations and qualitative discussions of phenomena in QED are especially facilitated by the use of Feynman diagrams. These diagrams define in a graphic form an algorithm by which the probability amplitude of a specific process is computed in perturbation theory. The lines on the diagrams represent the motion of particles and the vertices represent their interactions. Thus, for example, the diagrams in Figure 1 represent the scattering of a photon by an electron. The propagation of the photon is

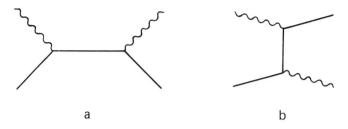

FIGURE 1. Feynman diagrams for photon-electron scattering.

shown here by wavy lines and that of the electron by straight lines. Lines with one free end correspond to free particles, colliding or emerging. A line joining two vertices corresponds to a so-called virtual particle for which $k^2 \neq m^2$ (here k is the 4-momentum of the particle and m is its mass; according to the Feynman rules, the interaction at each vertex conserves the 4-momentum).

In calculations, each virtual particle is put into correspondence with a function describing its propagation, the "propagator." In fact, it is the virtual particles that are responsible, in the framework of the diagram calculus, for the description of force fields through which interacting particles affect one another.

The virtual electron in Figure 1a carries a time-like momentum ($k^2 > m^2 > 0$). In Figure 1b, which also shows photon-electron scattering, the virtual electron can carry a space-like momentum ($k^2 < 0$). In Compton scattering, the force field is described by a virtual electron, but, in electron-electron scattering, the force field is described by a virtual photon (Figure 2).

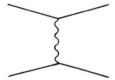

FIGURE 2. Feynman diagram for electron-electron scattering.

One spectacular property of Feynman diagrams is that their lines simultaneously describe the propagation of both particles (electrons) and antiparticles (positrons). The positron is interpreted as the electron propagating backward in time; that is to say, against the time arrow (the time arrow on diagrams is usually assumed to be directed from left to right).

The diagram in Figure 3 represents the annihilation of an electron and positron into two photons. The diagram in Figure 4 shows the reverse process: the creation of an electron-positron pair by two colliding photons. The diagram in Figure 5 represents the creation of a $\mu^+\mu^-$ pair in an electron-positron collision. All diagrams discussed up to now are examples of tree-type diagrams. The values of the 4-momenta of virtual particles in a tree diagram are determined by the values of the 4-momenta of real particles. Higher orders of perturbation theory produce loop diagrams (see, e.g., Figure 6) in which the momenta of virtual particles forming loops are not fixed and integration is carried out over them. The loop in Figure 6 is formed by a virtual electron-positron pair created by a virtual photon and later annihilating into a virtual photon. This formation of virtual pairs in the course of the photon's propagation in a vacuum is called vacuum polarization.

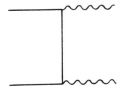

FIGURE 3. Feynman diagram for electron-positron annihilation into two photons.

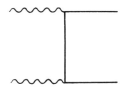

FIGURE 4. Feynman diagram for the creation of an electron-positron pair in the collision of two photons.

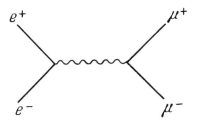

FIGURE 5. Feynman diagram for the annihilation of an electron-positron pair into a $\mu^+\mu^-$ pair.

GRAVITATION AND ELECTRODYNAMICS

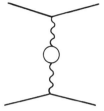

FIGURE 6. An example of a Feynman diagram with a vacuum polarization loop.

POLARIZATION OF THE VACUUM

In quantum electrodynamics, vacuum polarization results in the screening of the electric charge of electrons by vacuum positrons. By polarizing the vacuum, the electron attracts virtual vacuum positrons and repulses virtual vacuum electrons. As a result, if we observe the electron from a large distance, its charge becomes partially screened. If we penetrate deep into the cloud of virtual pairs however, the screening diminishes and the observed charge increases. The electric charge of the electron e is thus a function of distance: $e=e(r)$. The same is true, of course, for the quantity $\alpha(r)$, which for this reason is sometimes called the "running coupling constant." Since small distances r correspond to large transfers of momenta q $(r\sim\hbar/q)$, α is usually said to be a function of q. The standard value, $\alpha\approx 1/137$, refers to relatively large distances and small momentum transfers: $q \lesssim m_e c$. For $q \gg m_e c$, $\alpha(q)$ grows logarithmically as q increases.

We shall see below that the coupling constants of strong and weak interactions are also "running," but, in contrast to the electromagnetic constant, they do not increase but fall off as q increases. By extrapolating this "running," we notice that, at some sufficiently large momentum, the charges of all three interactions become identical. It is this observation that serves as the basis for the models of the grand unification of the electromagnetic, weak, and strong interactions (see Chapter VI, Figure 32).

Chapter 3

THE STRONG INTERACTION

Hadrons and quarks
Isotopic spin. SU(2) group
Strange particles
SU(3) symmetry
The charmed quark
The b quark and others
Flavors and generations
Color and gluons
Quantum chromodynamics (QCD)
Asymptotic freedom and confinement
Chiral symmetry
QCD on the march

HADRONS AND QUARKS

In contrast to leptons, hadrons can only be called elementary particles with certain reservations. All the numerous hadrons are indeed elementary, in the sense that none of them can be broken into constituents. At the same time, it has been reliably established that hadrons have an internal structure: they consist of quarks. Like leptons, quarks appear at our present level of knowledge as structureless, truly elementary particles. For this reason, leptons and quarks are often called "fundamental particles."

The paradoxical properties of quarks have no precedent in the history of physics, itself so rich in paradoxes. Experimentalists, using beams of high-energy particles, definitely observe these constituents inside hadrons and measure their spins, masses, and electric charges. But no one has succeeded—nor ever will succeed, if today's theoretical concepts are correct—in knocking a quark out of a hadron. Quarks in hadrons are imprisoned for life. Physicists use a milder term for this imprisonment—confinement. Theoretical concepts of the mechanism of confinement will be discussed later. Now we shall have a better look at different sorts of quarks.

It is convenient to begin our discussion of the properties of quarks with a nonrelativistic quark model that operates with the so-called constituent quarks. Hadrons are built out of the constituent quarks as if of building blocks. A constituent quark of a given type is a complex object with the same electric charge and the same spin as the "bare" quark present in the Lagrangian (these Lagrangian quarks are usually referred to as current quarks). The complex structure of the constituent quark appears because the current quark is surrounded by a cloud of virtual particles generated by strong interactions. As a result, the mass of a constituent quark exceeds that of the corresponding current quark by about 300 MeV. Hereafter, when we mention the masses of quarks, we shall always mean the masses of current quarks.

Protons and neutrons are built up of the lightest quarks, u (up) and d (down). As with all other quarks, their spin is 1/2. The charge of the u quark is $+2/3$, and that of the d quark is $-1/3$. The mass of the u quark is approximately 5 MeV and that of the d quark is 7 MeV. The proton is built of two u quarks and one d quark: p = uud. The neutron consists of two d quarks and one u quark: n = ddu.

According to the nonrelativistic quark model, quarks have zero angular orbital momenta within nucleons. The total spin of two u quarks in the proton is unity. This unity, added geometrically to the spin of the d quark, gives the proton a spin of 1/2. The neutron is constructed similarly, by interchanging u and d quarks.

A whole sequence of other hadrons can be constructed of quarks as of toy cubes. Thus, for example, if the spins of three quarks are parallel, they form the quartet of spin 3/2 Δ baryons:

$$\Delta^{++} = \text{uuu}, \Delta^{+} = \text{uud}, \Delta^{0} = \text{udd}, \Delta^{-} = \text{ddd}.$$

The orbital angular momenta of quarks in Δ-baryons, according to the nonrelativistic quark model, are zero. The attentive reader has probably noticed that this structure of Δ-baryons and nucleons seems to contradict the Pauli principle: indeed, two or even three quarks of the same sort seem to occupy the same quantum state. In fact, the Pauli principle is not in danger. We shall see below that these quarks differ from each other in color.

The Δ baryons are the lightest of baryon resonances. They decay into nucleons and π mesons in about 10^{-23} s: $\Delta \to N\pi$. A large number of heavier baryon resonances, also composed of u and d quarks, are known. In these resonances, quarks are in states with orbital and/or radial excitations. In this respect, resonances resemble excited states of atoms.

Baryons thus consist of three quarks. Another type of hadron—mesons—consists of a quark and an antiquark. For example, the lightest of mesons—π mesons—have the following structure:

$$\pi^+ = u\tilde{d}, \quad \pi^0 = \frac{1}{\sqrt{2}}(u\tilde{u} - d\tilde{d}), \quad \pi^- = d\tilde{u}$$

(the meaning of the minus sign in a quantum-mechanical superposition of states forming the π^0 meson will be explained later). The quark and antiquark in the π meson are in the state with zero orbital momentum and oppositely directed spins, so that the total spin of the π meson is zero.

If a quark and an antiquark with zero orbital momentum have parallel spins, then they form mesons with spin 1: ρ^+, ρ^0, ρ^-. These mesons are resonances and decay into two π mesons over a time of the order of 10^{-23} s: $\rho \to 2\pi$. The ρ mesons are the lightest among meson resonances. A large number of heavier resonances, in which the quark–antiquark pairs are in excited states, are known.

Decays of Δ and ρ resonances can be illustrated by the following quark diagrams. In Figures 7 and 8, an arrow directed backward in time represents an antiquark. Do not overlook the difference between conventional Feynman diagrams and quark diagrams: The quarks that fly to infinity are not free but confined within hadrons. Furthermore, as a rule, strong interactions between quarks are not indicated on quark diagrams. Thus, the interaction that results in the creation of a quark–antiquark pair, which appears as a "hairpin" on quark diagrams, is not shown.

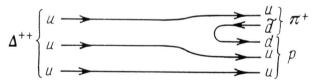

FIGURE 7. Quark diagram for the $\Delta^{++} \to p\pi^+$ decay.

FIGURE 8. One of the two quark diagrams for the $\rho^+ \to \pi^+\pi^0$ decay.

Figure 8 contains one of the two quark diagrams describing the decay of the ρ meson. I suggest that the reader construct the second diagram as an exercise.

ISOTOPIC SPIN. SU(2) GROUP

The difference between the masses of the u and d quarks is much smaller than the mass of a hadron composed of these quarks. Therefore, it is reasonable to consider an approximation in which the masses of the u and d quarks are equal. In the theory of the strong interaction that we shall describe shortly, strong interactions of the u and d quarks are identical. If the mass difference of the u and d quarks and the difference between their electric charges are neglected, then the quark Lagrangian has an additional symmetry called the isotopic symmetry.

In the framework of the isotopic symmetry, the u and d quarks are treated as upper and lower states of a spinor in the so-called isotopic space. The u quark corresponds to the projection of isotopic spin equal to $+1/2$ and the d quark to the projection of spin equal to $-1/2$. (The projection is onto some axis in the isospin space; it is usually referred to as the z axis.) Transformations of the isotopic spinor under which the Lagrangian is invariant are realized by complex 2×2 matrices (reads "two by two") satisfying the conditions of unitarity ($U^+U = I$, where U^+ is the Hermitian conjugate matrix, and I is the 2×2 identity matrix) and unimodularity (det $U = 1$). These 2×2 matrices give the simplest representation of the group SU(2) (reads "ess you two"). Here the letter S signifies that the transformations are special (unimodular in the present case) and the letter U indicates that they are unitary, while the numeral 2 denotes that the simplest representation of the group is formed by 2×2 matrices and that the simplest representation space is formed by a two-component spinor.

The group SU(2)—and the more complicated unitary unimodular groups SU(N), where $N>2$—play an important role in the physics of elementary particles. This calls for a more detailed discussion of the properties of 2×2 matrices U. Higher representations of SU(2) and representations of higher-order groups than SU(2) have much in common with these matrices.

In the general case, a 2×2 matrix U is determined by three real parameters α_k ($k = 1,2,3$) and can be written in the form

$$U = e^{i\alpha_k\tau_k/2} = 1 + i\alpha_k\frac{\tau_k}{2} + \frac{1}{2}\left(i\frac{\alpha_k\tau_k}{2}\right)^2 + \cdots$$

where summation is implied over the subscript k and τ_k are three Pauli matrices.

$$\tau_1 = \begin{pmatrix} 0 & 1 \\ 1 & 0 \end{pmatrix}, = \tau_2 = \begin{pmatrix} 0 & -i \\ i & 0 \end{pmatrix}, \text{ and } \tau_3 = \begin{pmatrix} 1 & 0 \\ 0 & -1 \end{pmatrix}.$$

When acting on a spinor, the matrix $\tau_+ = \frac{1}{2}(\tau_1 + i\tau_2)$ raises its lower component upward, while the matrix $\tau_- = \frac{1}{2}(\tau_1 - i\tau_2)$ lowers the upper component downward and the matrix $\frac{1}{2}\tau_3$ gives the eigenvalues of the projection of the isotopic spin on the z axis in the isotopic space. Pauli matrices do not commute:

$$[\tau_i, \tau_k] = \tau_i \tau_k - \tau_k \tau_i = i 2 \epsilon_{ikl} \tau_l \qquad (i, k, l = 1, 2, 3),$$

where ϵ_{ikl} is a completely antisymmetric tensor:

$$\epsilon_{123} = \epsilon_{231} = \epsilon_{312} = 1 \text{ and } \epsilon_{213} = \epsilon_{132} = \epsilon_{321} = -1.$$

Components of the tensor ϵ_{ikl} with at least two identical subscripts equal zero.

A group whose consecutive transformations do not commute with each other are called non-Abelian. The group SU(2) is one of the simplest non-Abelian groups.

SU(2) can be used to illustrate another important concept. If the parameters of group transformations (in this case, $\alpha_1, \alpha_2, \alpha_3$) do not depend on space-time coordinates, the symmetry is called global. But if they are functions of space-time coordinates, the symmetry is called local. In the second half of this chapter, we shall see that the isotopic symmetry caused by the resemblance of the properties of the u and d quarks is global; we shall consider a very interesting example of local symmetry involving the concept of color.

The foregoing mathematical definitions are presented for future use. They will help the reader understand the more complicated physical symmetries which will come up later both in this book and in other books on elementary particles. As for "constructing" baryons of three quarks or mesons out of a quark and an antiquark, elementary operations with such a "quark erector set" are intelligible to even the youngest schoolchildren. This is also true of a number of aspects of isotopic symmetry.

For instance, for an arbitrary isotopic multiplet with isotopic spin I, the number of particles in the multiplet is given by the simple formula

$$n = 2I + 1,$$

which is readily derivable from the observations that the maximum isospin projection on the third axis is I, the minimal projection is $-I$, and the "step" ΔI is unity. Correspondingly, the isospin of nucleons is 1/2, that of π mesons is 1, and that of Δ isobars is 3/2.

Note that this quark-based outline of isotopic symmetry is anti-historical. Historically, the concept of isotopic spin was introduced into physics by Heisenberg at the beginning of the 1930s, immediately after the discovery of the neutron, and was applied to nucleons and the nuclear forces between them. It was soon extended to the then-hypothetical π mesons, whose existence was predicted by Yukawa. Multiplets of real π mesons and Δ isobars were discovered approximately 20 years later. The quark hypothesis was suggested only in 1964. The path to this hypothesis lay through a study of the properties of the so-called strange particles and SU(3) symmetry.

STRANGE PARTICLES

The family of strange hadrons is more numerous than that of non-strange hadrons. In comparison to nucleons and π mesons, strange hadrons play a rather minor role in nuclear physics because strange hadrons are unstable (the most long-lived of them, the K_L^0 meson, lives for 5×10^{-8}s) and heavy, so that they are produced only at relatively high energies of colliding particles.

The first strange particles were discovered in cosmic rays in the 1940s. In the 1950s they were already being mass-produced by specially designed accelerators. What seemed paradoxical or strange in their behavior (and gave rise to the name) was that these particles are born copiously (strongly) (at sufficiently high energies of colliding hadrons), but decay into non-strange hadrons slowly (weakly). (On the nuclear scale of time, 10^{-8}s is a very long time, the characteristic time of the strong interaction being 10^{-23}s. Thus, the K_L^0 meson lives for about 10^{16} "nuclear days"—compare it to the age of the Earth: about 10^{12} terrestrial days.)

The solution to this paradox is that strange particles are produced in pairs via the strong interaction and decay individually via the weak interaction. At the present time we know that this happens because each strange particle contains at least one strange quark: the s quark.

Like the d quark, the strange quark's charge is 1/3. But it is much heavier than the d quark; its mass is about 150 MeV. Decays of s quarks will be discussed in the section devoted to the weak interaction. Here we shall look into their strong interactions. Strong interactions produce quark–antiquark pairs: s + s̄. Figure 9 shows a quark diagram of the process $\pi^-p \to K^0\Lambda^0$. Notice that the creation of a pair of strange particles results in a "strange hairpin" (ss̄) on the quark diagram. In this particular case, one end of the hairpin (s̄) belongs to a K meson and the other (s) to the Λ hyperon.

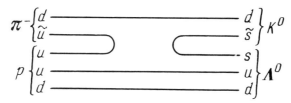

FIGURE 9. Quark diagram for the $\pi^-p \to K^0\Lambda^0$ reaction.

SU(3) SYMMETRY

The K meson is the lightest of the strange mesons and the Λ hyperon is the lightest of the strange baryons (strange baryons were given the name hyperons). Strange and non-strange hadrons together form common families: meson octets and singlets and baryon octets and decuplets (a singlet contains 1, an octet 8, and a decuplet 10 particles). The structure of these families is easily understood in terms of the SU(3) symmetry. In terms of quarks, the SU(3) symmetry is reduced to a symmetry (degeneracy) among u, d, and s quarks. This SU(3) symmetry is a generalization of the isotopic SU(2) symmetry.

The SU(3) symmetry is much more strongly broken in nature than SU(2) because the s quark is much heavier than non-strange quarks:

$$m_s - m_u \approx m_s - m_d \gg m_d, m_u$$

As a consequence, there are large mass splittings among hadrons within each SU(3) multiplet. It was not easy to deduce the existence of the SU(3) symmetry by studying hadrons. The decisive contribution to the understanding of symmetry properties of hadrons was made by Gell-Mann. At the beginning of the 1950s, he extended the notion of isotopic spin to strange particles. At the beginning of the 1960s, he gave the present formulation of the SU(3) symmetry of mesons and baryons. And finally, in 1964, he advanced the quark hypothesis (three independent contributions by Nishijima, Ne'eman, and Zweig should be mentioned in connection with the isotopic spin, SU(3), and quarks, respectively).

SU(3) multiplets are conveniently plotted on the plane $I_3 Y$ where I_3 is the third projection of isotopic spin, and Y is the hypercharge (by definition, hypercharge equals twice the mean charge of an isotopic multiplet). Figure 10 shows the octet of pseudoscalar mesons ($J^P = 0^-$, where J is the spin of the particles and P is their parity; parity will be discussed in detail in the chapter devoted to the weak interaction). Figure 11 shows the octet of vector mesons ($J^P = 1^-$). The quark structure of these SU(3) multiplets is given in Figure 12. The structure of the particles at the vertices of the hexagon is obvious, but the combinations at its center need clarification.

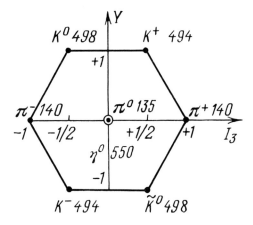

FIGURE 10. $I_3 Y$-diagram for the octet of the lightest pseudoscalar mesons.

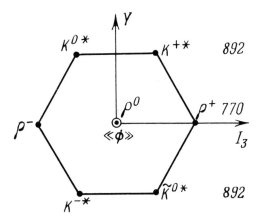

FIGURE 11. $I_3 Y$-diagram for the octet of the lightest vector mesons.

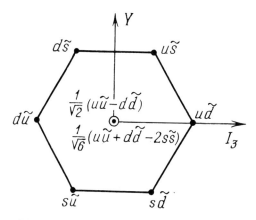

FIGURE 12. Quark structure of an octet of mesons.

Nine different combinations can be constructed out of three quarks and three antiquarks. Three of them are truly neutral: $u\tilde{u}$, $d\tilde{d}$, and $s\tilde{s}$. As a result of strong interactions, these three quark–antiquark states can transform into each other so that definite mass values exist for three quantum-mechanical superpositions of these states. Were the SU(3) symmetry strict, an SU(3)-invariant superposition,

$$\tfrac{1}{\sqrt{3}}(u\bar{u} + d\bar{d} + s\tilde{s}),$$

would split off. In the case of pseudoscalar mesons, this would correspond to the SU(3) singlet η' meson; for vector mesons, it would correspond to the SU(3) singlet ω meson. Among the two remaining superpositions, one has an isotopic spin equal to 1 (this is π^0 for pseudoscalars and ρ^0 for vectors),

$$\tfrac{1}{\sqrt{2}}(u\bar{u} - d\bar{d})$$

(it is constructed of quark wave functions by means of the matrix τ_3), and the other superposition has zero isospin:

$$\tfrac{1}{\sqrt{6}}(u\bar{u} + d\bar{d} - 2s\tilde{s}).$$

Its form is determined by requiring orthogonality to the first two superpositions. It corresponds to the η meson for pseudoscalars and the ϕ meson for vectors (note that the coefficients with all three superpositions stem from the normalization of the quantum-mechanical state to unity).

Like the three Pauli matrices τ in the case of SU(2) symmetry, the matrices important in SU(3) are the eight Gell-Mann matrices λ:

$$\lambda_1 = \begin{pmatrix} 0 & 1 & 0 \\ 1 & 0 & 0 \\ 0 & 0 & 0 \end{pmatrix}, \lambda_2 = \begin{pmatrix} 0 & -i & 0 \\ i & 0 & 0 \\ 0 & 0 & 0 \end{pmatrix}, \lambda_3 = \begin{pmatrix} 1 & 0 & 0 \\ 0 & -1 & 0 \\ 0 & 0 & 0 \end{pmatrix}, \lambda_4 = \begin{pmatrix} 0 & 0 & 1 \\ 0 & 0 & 0 \\ 1 & 0 & 0 \end{pmatrix},$$

$$\lambda_5 = \begin{pmatrix} 0 & 0 & -i \\ 0 & 0 & 0 \\ i & 0 & 0 \end{pmatrix}, \lambda_6 = \begin{pmatrix} 0 & 0 & 0 \\ 0 & 0 & 1 \\ 0 & 1 & 0 \end{pmatrix}, \lambda_7 = \begin{pmatrix} 0 & 0 & 0 \\ 0 & 0 & -i \\ 0 & i & 0 \end{pmatrix}, \lambda_8 = \frac{1}{\sqrt{3}} \begin{pmatrix} 1 & 0 & 0 \\ 0 & 1 & 0 \\ 0 & 0 & -2 \end{pmatrix}.$$

One readily notices a relationship between the quark structure of the η meson and the matrix λ_8.

Since the SU(3) symmetry is broken in nature, SU(3) singlet mesons and the eighth components of SU(3) octets are partly mixed. This phenomenon is called mixing. Mixing is stronger for vector mesons than for pseudoscalar mesons. The physical states produced by this mixing are

$$\omega \approx \tfrac{1}{\sqrt{2}}(\bar{u}u + \bar{d}d) , \quad m = 783 \text{ MeV},$$

$$\phi \approx s\bar{s} \quad , \quad m = 1020 \text{ MeV}.$$

Figure 13 shows the baryon octet with $J^P = \tfrac{1}{2}^+$. The quark structure is given in simplified form in Figure 14. The combination $|ud|s$ at the center of Figure 14, symmetric with respect to the substitution u↔d, has isospin 1 and describes the Σ^0 hyperon; the combination [ud]s, antisymmetric with respect to u↔d, has $I = 0$ and describes the isoscalar Λ^0 hyperon. Figures

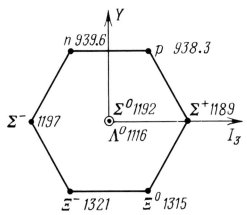

FIGURE 13. I_3Y-diagram for the octet of the lightest baryons with $J^P = 1/2^+$.

15 and 16 show the baryon decuplet with $J^P = 3^+/2$ and its quark structure. A number of SU(3) multiplets with other values of spin and parity are known. But the decisive role in establishing the SU(3) symmetry and the quark structure of hadrons was played by the octets shown in Figures 10 and 13 and the decuplet shown in Figure 15.

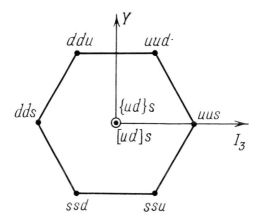

FIGURE 14. Quark structure of an octet of baryons.

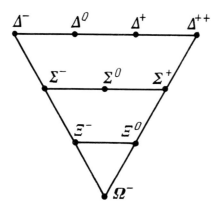

FIGURE 15. Decuplet of the lightest baryons with $J^P = 3/2^+$.

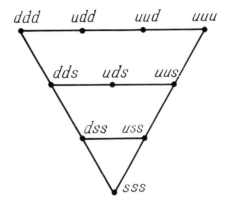

FIGURE 16. Quark structure of a decuplet of baryons.

THE CHARMED QUARK

The quark theory of hadrons was conclusively confirmed by the discovery of charm made by Richter and Ting and their collaborators.

Charmed particles comprise quarks of the fourth sort, the so-called charmed quarks denoted by the letter c. First came the discovery in the autumn of 1974 of the J/ψ meson, a vector particle with "hidden charm" consisting of a pair c$\bar{\text{c}}$ in the $^3S_1$ state. A number of other levels of the c$\bar{\text{c}}$ system (dubbed charmonium) were soon discovered. The diagram of the charmonium levels currently known is shown in Figure 17. The masses of the levels are given in GeV (the diagram is purely schematic, with no mass scale). Primed particles represent radial excitations of lower states. S-states correspond to zero orbital momentum of c$\bar{\text{c}}$; P-states to an orbital momentum equal to unity. The right-hand lower suffix indicates the spin of the meson and the upper left-hand suffix indicates the total spin of the quark and antiquark.

Some particles with explicit charm were also found: the mesons D^0 [c$\bar{\text{u}}$] (1.863), D^+ [c$\bar{\text{d}}$] (1.868), and F^+ [c$\bar{\text{s}}$] (2.02) and the baryon Λ_c^+ [cdu] (2.27) (the symbols in brackets are the quark compositions of the particles; the numbers in parentheses are the masses of the particles expressed in GeV).

The study of the properties of these particles made it possible to determine not only the charge but also the mass of the c quark. The charge of the c quark is $+2/3$ and its mass is approximately 1.4 GeV. Therefore, the c quark is a very heavy analogue of the u quark.

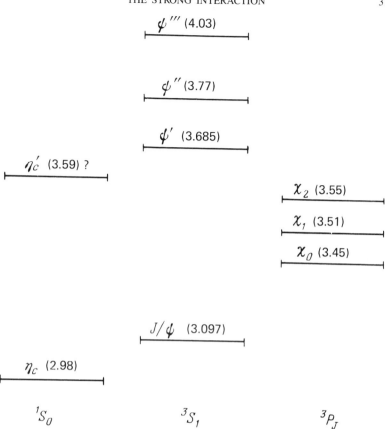

FIGURE 17. Spectroscopy of charmonium states. The masses of the states (energy levels) are given in GeV in parentheses; vertical spacings between levels in the figure are arbitrary.

THE b QUARK AND OTHERS

In 1976, Lederman and his collaborators at FNAL discovered a new particle: the Y meson (read "upsilon"), consisting of quarks of the fifth sort, the so-called b quarks. The charge of the b quark was found to be $-1/3$. The b quark is a heavy analogue of the "lower" d and s quarks; hence, the term "bottom" and the symbol b. (Some physicists prefer to derive the letter b

from beauty.) The b quark is much heavier than the c quark: its mass is approximately 4.8 GeV. The Y meson, whose mass is 9.46 GeV, is the lowest $^3S_1$ state of the hidden-beauty pair $b\bar{b}$. Three radially excited $^3S_1$ levels of this system, sometimes called upsilonium (and sometimes bottomonium), have been found: Y' (10.02), Y" (10.40), and Y''' (10.55).†
Mesons with a single b quark have also been found: $B^+ = \bar{b}u$, $B^0 = \bar{b}d$, $B^- = b\bar{u}$, and $\bar{B}^0 = b\bar{d}$. The masses of these naked-beauty mesons (~5.274 GeV) are such that

$$m_{Y''} < 2m_B < m_{Y'''}.$$

The existence has thus been established of three quarks of the "lower type": d,s,b.; and two quarks of the "upper type": u,c. It is expected, owing to quite persuasive arguments (see below), that a third "upper" quark also exists. It is called the t quark (from "top"). The search for the t quark has so far been unsuccessful. The search for "toponium," the $\bar{t}t$ pair, in colliding electron–positron beams at PETRA at an energy of 18 GeV in each beam has been especially thorough. It has been concluded from these experiments that if the t quark exists, its mass must be greater than 18 GeV. At present, there are no serious arguments in favor of quarks still heavier than t.

FLAVORS AND GENERATIONS

Quarks of different types are often said to differ by their flavors. These quark flavors have nothing in common with the ordinary notion of flavors. The word "flavor" is used here as a synonym of the words "type" or "sort," adorning by its unexpected appearance "dry" texts in physics. The term "flavor" is also convenient because it is semantically in contrast to the term "color" that we shall start discussing in the next section.

There seems to exist some profound symmetry between leptons and quarks of different flavors. The following table points to the existence of such a symmetry:

ν_e	ν_μ	ν_τ
e	μ	τ
u	c	$t(?)$
d	s	b

---

†*Author's note (autumn 1983)*. P-levels of upsilonium have recently been discovered.

It was the quark-lepton symmetry that allowed the existence of the c quark to be predicted as early as 1964 (four leptons and only three quarks were known at that time). After the τ lepton was discovered in 1975, the same symmetry was used to predict the existence of b and t quarks.

We shall soon see that the lepton-quark symmetry is especially well pronounced in weak interactions. Of course, this symmetry is not perfect: Although the differences between the charges of the neutrino and charged leptons equal those between the charges of the upper and lower quarks, the charges themselves are different for leptons and quarks.

The twelve leptons and quarks are naturally classified into three groups, or three generations of fundamental fermions. Each generation consists of four particles forming a column in the table: an "upper" and a "lower" lepton and "upper" and "lower" quark. The first generation is formed of the lightest particles. In each subsequent generation charged particles are heavier than in the preceding one.

Together with photons, fermions of the first generation form the matter of which the universe is built at present. Atomic nuclei consist of u and d quarks, atomic shells consist of electrons, and nuclear fusion reactions inside the sun and stars would be impossible without the emission of electron neutrinos. As for the fermions of the second and third generations, their role in the world around us appears to be negligible. At first glance, the world would not seem to be any worse off if these particles never existed. These particles resemble draft versions that the Creator has thrown out as unsuccessful, but that we, using sophisticated instruments, have retrieved from his waste basket.

Now we have begun to understand that these particles were very important at the first instants of the Big Bang. Thus, the number of sorts (flavors) of neutrinos determined the ratio between the hydrogen and helium abundances in the universe. Cosmological calculations of helium abundance indicate that the number of neutrino flavors does not exceed four. In the framework of the scheme of lepton-quark generations, this means that the total number of quark flavors is not greater than eight.

The importance of the subsequent generations seems to lie in that it is because of them that the particles of the first generation have precisely the values of mass that we observe. The relation between the masses of the u and d quarks and the electron is essential for our very existence. Indeed, the difference between the masses of the neutron and proton is primarily due to the difference between the masses of the u and d quarks. And if the inequality $m_p - m_n + m_e > 0$ were correct, hydrogen would be unstable.

We thus begin to guess that higher-order generations are not as insignificant as we thought. To find their profound significance, and the nature of the quark-lepton symmetry itself, is one of the most important problems in physics. These remarks conclude the discussion of quark flavors and we shall now begin a new topic: quark colors.

COLOR AND GLUONS

So far we have carefully avoided the question of how the forces between quarks are realized. It is time to answer at this juncture the following questions: (i) What charges are the sources of these forces? and (ii) What particles mediate these forces? Short answers to these two questions are: (i) color charges and (ii) gluons.

It has been established that quarks of each flavor exist in three strictly degenerate species. These species are said to differ only in color. The following three colors are normally used: yellow, blue, and red. These quark colors obviously have no bearing on the usual optical colors. "Color" for quarks is simply a convenient term to denote a quantum number that characterizes quarks. Color charges of antiquarks are different from those of quarks. Sometimes they are referred to as antiyellow, antiblue, and antired, and sometimes as violet, orange, and green, in correspondence with the well-known sequence of complementary colors in the optical spectrum (recall the convenient mnemonic phrase "Richard Of York Gained Battles (In) Vain").

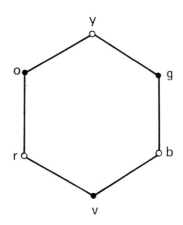

FIGURE 18. Three colors of quarks (red, blue, yellow) and three colors of antiquarks (green, orange, violet).

THE STRONG INTERACTION

With such a selection of quark colors it is natural to regard hadrons as colorless, or white, particles. Baryons are colorless because they consist of three quarks of three mutually complementary colors. Mesons are colorless superpositions of quarks and antiquarks.

Mathematically, color degeneracy of quarks implies the color SU(3) symmetry: $SU(3)_c$ (subscript c for color). A color triplet of quarks, q^α, $\alpha = 1,2,3$, for yellow, blue, and red forms the so-called fundamental representation of the group SU(3). A triplet of antiquarks, $\tilde{q}_\alpha$, forms a conjugate representation (this is an antitriplet). Mesons (M) and baryons (B) are $SU(3)_c$ singlets:

$$M = \frac{1}{\sqrt{3}} \tilde{q}_\alpha q^\alpha = \frac{1}{\sqrt{3}} (\tilde{q}_1 q^1 + \tilde{q}_2 q^2 + \tilde{q}_3 q^3),$$

$$B = \frac{1}{\sqrt{6}} q^\alpha q^\beta q^\gamma \epsilon_{\alpha\beta\gamma},$$

where $\epsilon_{\alpha\beta\gamma}$ is a completely antisymmetric tensor that we have encountered already when discussing the properties of the Paul matrices. It is because of the antisymmetrization with respect to color that three quarks in a baryon do not violate the Pauli principle and thus behave as ordinary fermions.

The color charges of quarks play the same part in the strong interaction that the electric charges of particles play in the electromagnetic interaction; and the photon's part is played by electrically neutral vector particles christened gluons. Quarks exchanging gluons are thus "glued together" to form hadrons.

The main difference between gluons and photons is that there is only one photon and it is electrically neutral, but there are 8 gluons and they all carry color charges. Because of their color charges, gluons strongly interact with one another and emit one another. As a result of this nonlinear interaction, the propagation of gluons through vacuum is quite unlike the propagation of photons, and color forces do not resemble electromagnetic forces.

When a beam of light traverses a dark room it is visible to a side observer only because light is scattered by suspended dust. Otherwise it would be invisible since photons are neutral and therefore do not emit photons. Gluons do emit gluons, and one would expect that a beam of gluons, in a "dark" room, would behave as a "shining light." As we shall shortly see, this is not so because of a phenomenon called confinement.

QUANTUM CHROMODYNAMICS (QCD)

The theory of the interaction between quarks and gluons is called quantum chromodynamics (from the Greek word $\chi\rho\omega\mu\alpha$, color). QCD is based on the postulate that the $SU(3)$ color symmetry is a local, i.e., gauge symmetry. The requirement of local invariance (gauge invariance) leads to the inevitable conclusion of the existence of the octet of gluon fields with their specific self-coupling. Thus, the symmetry determines the entire dynamics of strong interactions. In this respect, the local color symmetry $SU(3)_c$ is much deeper than the global flavor symmetry, $SU(3)_f$ (subscript f for flavor), which appears because the masses of the u, d, and s quarks are approximately degenerate.

The Lagrangian of QCD is very much like the QED Lagrangian (see Chapter II). The difference lies in that (i) the electromagnetic coupling constant, that is, electric charge e, is replaced by the strong coupling constant g, (ii) quark spinors, in contrast to the electron spinor, have color suffices over which summation is carried out, and (iii) the gluon vector potential A_μ in the Lagrangian, in contrast to the photon Lagrangian, is a matrix in color space:

$$A_\mu = A_\mu^i \lambda_i/2, \quad i = 1,2,\ldots, 8.$$

Here A_μ^i are vector potentials of eight gluon fields and λ^i are eight Gell-Mann matrices. Note that, in QCD, the covariant derivative takes the following (matrix) form:

$$D_\mu = \partial_\mu - igA_\mu.$$

The matrix of the gluon field strength is

$$F_{\mu\nu} = F_{\mu\nu}^i \lambda_i/2.$$

In the case of gluons, the field strength $F_{\mu\nu}$ is expressed through A_μ by a more complex formula than in the case of photons:

$$F_{\mu\nu} = \partial_\mu A_\nu - \partial_\nu A_\mu - ig[A_\mu A_\nu - A_\nu A_\mu].$$

Here g is the strong interaction constant. For photons, A_μ is a number, not a matrix, and the commutator in the expression for $F_{\mu\nu}$ vanishes. But, for non-Abelian gauge fields, such as gluons, this commutator is not equal to zero. Consequently, it determines the character of the nonlinear self-interaction of gluons peculiar to gluon forces.

This form of the QCD Lagrangian, and of the tensor $F_{\mu\nu}$ is dictated by the requirement that the Lagrangian be invariant under gauge transformation:

$$q \to Sq, \quad \bar{q} \to \bar{q}S^+, \quad A_\mu \to SA_\mu S^+ - \frac{i}{g}(\partial_\mu S)S^+$$

where $S = \exp[-i\alpha_i(x)\lambda_i/2]$ and α_i are eight parameters that depend on the coordinates of the world point x.

ASYMPTOTIC FREEDOM AND CONFINEMENT

If we take into account the contribution of the nonlinear coupling of gluons to the polarization of the gluon vacuum by a gluon (see Figure 19), we find that, in contrast to the polarization of the quark vacuum by a gluon (Figure 20), this polarization results not in a screening but in an antiscreening of the color charge. The color charge of a quark diminishes as we go deeper into the gluon cloud surrounding the quark. This means that, in the limit of infinitely small distances between quarks, the color interaction between them "switches off." This phenomenon was given the name "asymptotic freedom." Quarks are nearly free at short distances: the color potential between them is a Coulomb-type potential, α_s/r, where the running constant $\alpha_s(r) = g^2(r)/4\pi$ decreases logarithmically as r diminishes or as the momentum transfer q increases. At sufficiently large q,

$$\alpha_s(q) \approx \frac{2\pi}{b \ln(q/\Lambda)}.$$

Here the dimensionless coefficient b is found theoretically by calculating the diagrams in Figures 19 and 20; $b = 11 - (2/3)n_f$, where n_f is the number of quark flavors ($b = 7$, if $n_f = 6$).

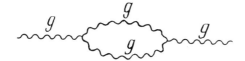

FIGURE 19. Gluon loop: the contribution to the polarization of vacuum by a gluon.

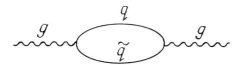

FIGURE 20. Quark loop: the contribution to the polarization of vacuum by a gluon.

As for the constant Λ, which has the dimension of momentum, its value is extracted from experimental data (using the widths and masses of energy levels of heavy quarkonia, the properties of hadronic jets produced in e^+e^- annihilations at high energies, and the properties of cross section of deep inelastic scattering) and is found to be of the order of 0.1 GeV. The constant Λ (sometimes denoted Λ_{QCD}) plays a fundamental role in QCD.

The reverse side of asymptotic freedom is the growth of the color charges as the distance between quarks increases. At distances $r \sim 1/\Lambda \sim 10^{-13}$ cm, color interaction becomes truly strong. In this region, perturbation theory fails and no reliable calculations can be made. Nevertheless, qualitative arguments show that the strengthening of this interaction with distance can be expected to result in quark confinement, that is, the impossibility of getting single free quarks.

In order to clarify the anticipated picture of confinement, imagine first a world completely devoid of light quarks. Consider a heavy quark and a heavy antiquark ($m \gg \Lambda$). At short distances ($r \ll 1/\Lambda$), the color potential between the quarks resembles the Coulomb law ($\sim 1/r$), and the force decreases with distance as $\sim 1/r^2$. Lines of color force diverge from the charge isotropically, so that their flux across a unit surface area is inversely proportional to the whole surface area (Figure 21). At large distances between quarks ($r \gg 1/\Lambda$), the strong nonlinear interaction between gluons makes the surrounding vacuum "squeeze" the lines of force into a tube with the radius $\sim 1/\Lambda$. This produces a "gluonguide" resembling a conventional lightguide. In this situation, the flux across a unit surface area is constant, the force between quarks is independent of the distance between them, and the potential is proportional to this distance. As a result, the color potential resembles a funnel (Figure 23).

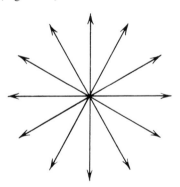

FIGURE 21. Lines of force of the Coulomb field.

FIGURE 22. Lines of force of the gluon field between a quark and an antiquark.

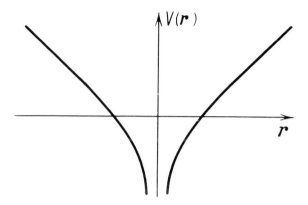

FIGURE 23. Funnel-like potential between a quark and an antiquark.

The levels of charmonium and upsilonium calculated on the basis of the funnel-type phenomenological potential are in good agreement with experimental data. However, it has not been possible thus far to construct an analytic theory of the gluon string based on a solution of the equations obtained from the QCD Lagrangian. But a large number of numerical computations on powerful computers, with QCD equations simplified by converting the space-time continuum to four-dimensional lattices with a finite number of cells (up to 10^4), point to the existence of such strings.

In principle, a gluon string between a heavy quark and a heavy antiquark could be infinitely extended. The energy spent on pulling the quarks apart would thereby be transformed into the mass of the string. However, in the real world where light quarks exist ($m \ll \Lambda$), this does not occur. The string breaks into segments about $1/\Lambda \sim 10^{-13}$ cm long, constituting new mesons. A light quark-antiquark pair appears at the breaking point (see Figure 24).

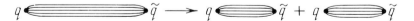

FIGURE 24. Gluon strong broken because of the production of a quark–antiquark pair. As a result, one meston transforms into two.

An attempt to "break" a meson into a quark and an antiquark would be similar to trying to break a magnetic needle into the south and north poles. We would always get two dipoles.

CHIRAL SYMMETRY

Since quarks are dressed with dense virtual gluon and quark–antiquark clouds, one cannot specify the mass of the quark without qualifying the distance at which this mass is measured. The shorter this distance, the smaller the mass is.

The masses given in the preceding sections refer to distances of the order of 10^{-14} cm, at which the clouds are rarified owing to asymptotic freedom. It has already been mentioned that these "half-bare" quarks devoid of their heavy gluon coat are usually called current quarks; quarks completely enveloped by gluon clouds are called block or constituent quarks. Assuming that the mass of a nucleon is the sum of the masses of the three nonrelativistic constituent quarks, we conclude that the masses of the gluon clouds around the u and d quarks are approximately 300 MeV.

It will be interesting to consider an imaginary world in which the current masses of light quarks are zero. It can be expected that this massless limiting case would fit the real world quite closely since the current masses of the u and d quarks are small: $m_u \approx 5$ MeV and $m_d \approx 7$ MeV. Indeed, a theoretical analysis shows that the masses of all baryons and nearly all mesons remain practically unaltered. The only exceptions are the lightest among mesons: π mesons whose squared masses are proportional to $(m_u+m_d)\Lambda_{QCD}$. This special position of the π mesons stems from the key role they play in the spontaneous breaking of chiral symmetry. We shall explain now what stands behind the terms "chiral symmetry" and "spontaneous breaking" of a symmetry.

By analyzing the QCD Lagrangian with massless u and d quarks, it is easy to notice that it has not only the isotopic symmetry SU(2) but also a higher global symmetry $SU(2)_L \times SU(2)_R$. The point is that massless particles possess a specific conserved characteristic that cannot be defined in a Lorentz-invariant manner for particles with nonzero mass. This quantity is the so-called helicity, that is, the projection of the spin of a particle on its momentum. Helicity is said to be left-handed (L) if the directions of spin and momentum are opposite and right-handed (R) if they are identically oriented. Massless particles move at the velocity of light; thus, it is impossible to change the helicity of a massless particle by any motion of a reference frame, whose velocity is always less than c. However, this is easy for a particle with nonzero mass.

Emission and absorption of vector gluons by color charges of quarks do not change quark helicities. Hence, the QCD Lagrangian of massless quarks naturally decomposes into two symmetric terms, one of which contains left-handed quarks u_L, d_L and the other, right-handed quarks u_R, d_R. Each of these terms is isotopically symmetric, so that the total Lagrangian is invariant under the product $SU(2)_L \times SU(2)_R$.

At the level of the Lagrangian, both symmetries $SU(2)_L$ and $SU(2)_R$ are on the same footing as the ordinary isotopic symmetry $SU(2)$. However, the difference becomes obvious when we look into how these symmetries are realized in the world of hadrons.

The ordinary isotopic symmetry is realized linearly: rotations of the isospinor of quarks and the isospinor of composite nucleons occur synchronously. This is not the case for chiral symmetry because, in contrast to massless quarks, nucleons are massive and have no definite helicity. Here, for the first time, we encounter a case in which the Lagrangian has a definite symmetry but the physical states have not. In such situations, we speak of a spontaneous breaking of symmetry. In this particular case, we face an example of spontaneous breaking of global chiral symmetry.

It was established that spontaneous symmetry breaking is always accompanied by the appearance of massless zero-spin bosons, the so-called Goldstone bosons. The three massless π mesons in our imaginary world in which the u and d quarks are massless would represent such Goldstone bosons. In the real world, where the masses of the u and d quarks are small but distinct from zero the chiral symmetry of the Lagrangian is approximate, the π mesons are called pseudo-Goldstone bosons: although their masses are not zero, they are small compared to the masses of other hadrons.

In principle, in the chiral limit, all the masses of hadrons consisting of light quarks should be expressible through a single dimensional parameter Λ_{QCD} that enters the expression for the running constant α_s. This problem has not yet been solved.

QCD ON THE MARCH

The advent of quantum chromodynamics drastically changed the situation in elementary particles physics. QCD explained the fundamentals of such already familiar symmetries as isotopic invariance $SU(2)$ and its generalization, the flavor SU(3) symmetry of strong interactions, and the chiral symmetries $SU(2)_L \times SU(2)_R$ and $SU(3)_L \times SU(3)_R$. New light was shed on such phenomenological models as the model of nonrelativistic quarks, the bag model, and the parton model. A number of new physical objects and phenomena were predicted on the basis of quantum chromodynamics: quark and gluon jets and glueballs (hadrons consisting only of gluons, without quarks).

No competitors contest the role of quantum chromodynamics as the ultimate theory of strong interactions. The main mountain pass on the way to a total understanding of hadrons is behind us: the Lagrangian has been written. Nevertheless, we are still far from the final goal because we are unable to solve the QCD equations as soon as the color interaction becomes strong. The problem of confinement is a challenge to the theorists. The mathematical structure of the theory and, in particular, the properties of the chromodynamic vacuum, with its quark and gluon condensates and intricate topological relief whose simplest elements are the so-called instantons, remains largely unprobed.

Further experimental study of hadrons may prove invaluable to further progress of the theory. It is remarkable that not only experiments at the highest possible energies but also experiments at low energies will be valuable for the theory. The latter will make it possible to put in order the spectroscopy of hadrons, including the spectroscopy of exotic (not of the types $\bar{q}q$ and qqq) and cryptoexotic mesons and baryons, baryonium, two-baryon resonances, and glueballs. (We advise the reader frightened by the avalanche of new terms to look them up in the Glossary.)

When Yang and Mills published, in 1954, a paper that pioneered the analysis of the local non-Abelian theory (SU(2)), it was difficult to discern in it the prototype of the future theory of strong interactions. Indeed, the theory operated with massless gauge fields that seemed to lead inevitably to long-range forces that do not exist in nature. Many a theorist considered the Yang-Mills theory an exciting mathematical toy. A long development that culminated in quarks proved necessary before Nambu could introduce, in 1965, a hypothesis of gauge fields coupled to the degree of freedom that Gell-Mann later (at the beginning of the 1970s) named color.

But QCD is not the only descendant of the Yang–Mills theory. The reader will see later that the current theory of the electroweak interaction and the models of grand unification of the strong, weak, and electromagnetic interactions are also realized as non-Abelian gauge theories.

Chapter 4

THE WEAK INTERACTION

Weak decays
Weak reactions
Components of the charged current
Mirror asymmetry
V −A current
C,P,T symmetries
Neutral currents
Neutrino masses and oscillations. Double β-decay
On the reliability of experimental data

WEAK DECAYS

February 1996 will mark the centennial of Becquerel's discovery that uranium salts emit a penetrating radiation. At that time Becquerel did not know, as we now do, that the rays he observed were the so-called β-rays; namely, electrons emitted in the radioactive decay of thorium. This was the discovery of β-decay and the beginning of the study of the weak interaction. (Note that the discovery of radioactivity also marks the beginning of the study of nuclear forces. We can thus say that the strong and weak interactions have a joint birthday.)

The first period in the history of the β-decay was completed at the beginning of the 1930s when Pauli, forced by a wealth of experimental data, put forward a hypothesis that β-decaying nuclei emit, together with electrons, light neutral particles: neutrinos. Soon after this, Fermi published a quantum-field theoretical description of β-decay. According to this theory, the neutron decay results from the interaction of two currents. One current, which today we would call the hadronic current, converts a neutron into a proton. Another current, the leptonic current, creates an electron-antineutrino pair. The interaction of these currents was called the four-fermion interaction because it involves four fermions.

The four-fermion coupling constant, or Fermi constant, is dimensional: $G_F = 1.436 \times 10^{-49}$ erg cm^3. In $\hbar = c = 1$ units, $G_F \approx 10^{-5} m_p^{-2}$, where m_p is the proton mass. On the nuclear scale, the Fermi constant is small; hence, the probability of the β-decay process, which is proportional to G_F^2, is also small.

After muons, π mesons and, especially, strange hadrons were discovered, it was found that the decays of all these particles, like the β-decay of nuclei, are caused by the weak four-fermion interaction with constant G_F. A very wide spread in lifetimes (e.g., the muon lives about $2\mu s$ while the neutron lives about 1000s) is naturally explained by the differences in energy Δ released in the decay, because decay rates are proportional to $G_F^2 \Delta^5$. It was thus established that the weak interaction is responsible for all slow decays. Subsequent studies of new types of particles (charmed particles, the τ lepton, and B mesons) has confirmed this universal character of the weak interaction. Among other things, the approximate relation $G_F^2 \Delta^5$ relation for decay rates was reliably borne out by experiments. Thus, for example, the τ lepton and charmed mesons are approximately twenty times as heavy as the muon. Correspondingly, their lifetimes are shorter by approximately seven orders of magnitude: around 10^{-13} s.

The ēν and ñp currents belong to the class of so-called charged currents. This term is used in the physics literature to supplant a more cumbersome but possibly more understandable term, "currents that change the electric charges of the particles involved." In both currents, the charge is reduced by 1: a neutral neutrino transforms into a negatively charged electron and a proton transforms into a neutron. In this interpretation, we take into account the facts that the operator ν annihilates a neutrino and that the operator ē creates an electron (and likewise for nucleons). But the operator ν not only annihilates the neutrino, but can also create an antineutrino, so that it can be said that the negative current ēν creates a negatively charged pair: electron + antineutrino. The same current annihilates a positively charged pair: positron + neutrino.

Along with the currents ēν and ñp, there are conjugate, positive currents ν̄e and p̄n, which increase the electric charge of the particles involved. These currents create positively charged pairs and annihilate negatively charged pairs of fermions.

Obviously, the β-decay interaction conserves electric charge. Correspondingly, its Lagrangian is a product of the positively charged current p̄n and the negatively charged current ēν.

WEAK REACTIONS

The coupling of the currents ēν and p̄n postulated by Fermi as the cause of the neutron β-decay n→pe$\tilde{\nu}_e$ (see Figure 25) must also lead to the reaction ν_en→pe$^-$ (see Figure 26). Indeed, we already know that the same operator realizes the creation of an antineutrino and the annihilation of a neutrino.

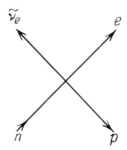

FIGURE 25. The decay of the neutron: $n{\rightarrow}pe\tilde{\nu}_e$.

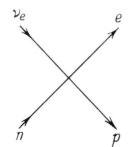

FIGURE 26. The $\nu_e n{\rightarrow}pe$ reaction.

Likewise, a product of conjugate currents ($\tilde{\nu}_e$e) and (ñp) gives the proton decay reaction p → ne$^+\nu_e$ (this occurs in certain nuclei in which the proton binding energy is smaller than that of the neutron) and the reaction $\tilde{\nu}_e$p → ne$^+$.

The reaction $\tilde{\nu}_e$p → ne$^+$ was first observed only in 1956, by using a beam of antineutrinos emitted by a nuclear reactor. This experiment, realized by Reines et al., marked the beginning of the study of weak reactions (before that, only weak decays had been observed experimentally).

In 1962, the first successful accelerator neutrino experiment, in which inelastic collisions of neutrinos with atomic nuclei were observed, was carried out at the Brookhaven National Laboratory (USA). The neutrino beam was generated by decaying π mesons ($\pi^+{\rightarrow}\mu^+\nu_\mu$ and $\pi^-{\rightarrow}\mu^-\tilde{\nu}_\mu$), which, in turn, were created by a proton beam colliding with a target. This experiment established that the muon and electron neutrinos are indeed different particles.

Weak nuclear forces were observed for the first time in 1964 at the Institute of Theoretical and Experimental Physics in Moscow, in a nuclear reaction produced by a beam of neutrons from a nuclear reactor. In Fermi's terms, these are forces due to the coupling of the current $\bar{p}n$ to a conjugate current $\bar{n}p$. This discovery confirmed the hypothesis, suggested much earlier, that the universal weak charged current interacts with its conjugate current.

If the whole charged current consisted only of two terms ($\bar{p}n + \bar{\nu}_e e$), the product ($\bar{p}n + \bar{\nu}_e e$)($\bar{n}p + e^-\nu_e$) would have four terms. The electron β-decay reveals the term ($\bar{p}n$)($\bar{e}\nu_e$). The positron β-decay reveals the conjugate term ($\bar{\nu}_e e$)($\bar{n}p$). Weak nuclear forces manifest the diagnonal term ($\bar{p}n$)($\bar{n}p$). The other diagonal term ($\bar{\nu}_e e$)($\bar{e}\nu_e$) must manifest itself in the scattering of electron neutrinos by electrons. This process was first observed experimentally only in 1976, but by that time it was beyond doubt that the current × current scheme was correct. Only the current has changed: nucleons were replaced with quarks, and the number of leptons has tripled.

COMPONENTS OF THE CHARGED CURRENT

All we know today about weak decays and reactions due to charged currents can be expressed as the result of the coupling of the total charged current j to its conjugate current j^+. The current j was shown to have nine components: three leptonic ($\bar{e}\nu_e$, $\bar{\mu}\nu_\mu$, $\bar{\tau}\nu_\tau$) and six quark components ($\bar{d}u$, $\bar{s}u$, $\bar{b}u$, $\bar{d}c$, $\bar{s}c$, $\bar{b}c$). Assuming also the existence of the sixth quark t, we have to add to the current j three more components ($\bar{d}t$, $\bar{s}t$, $\bar{b}t$). Hereafter, we shall discuss the properties of the weak current assuming that the t quark exists and writing j in the form:

$$j = \bar{e}\nu_e + \bar{\mu}\nu_\mu + \bar{\tau}\nu_\tau + a_{\bar{d}u}\bar{d}u + a_{\bar{s}u}\bar{s}u + a_{\bar{b}u}\bar{b}u \\ + a_{\bar{d}c}\bar{d}c + a_{\bar{s}c}\bar{s}c + a_{\bar{b}c}\bar{b}c + a_{\bar{d}t}\bar{d}t + a_{\bar{s}t}\bar{s}t + a_{\bar{b}t}\bar{b}t,$$

where $a_{\bar{d}u}$, $a_{\bar{s}u}$, ..., $a_{\bar{b}t}$ are numerical coefficients.

Even a cursory glance at this expression will show a salient difference between leptons and quarks. First, leptons form currents only with "their" neutrinos, while all the "upper" quarks form currents with each of the "lower" quarks, regardless of the generation to which they belong. Second, the three leptonic pairs enter the Lagrangian with coefficients equal to unity; this means that their couplings are absolutely identical. At the same time, all the coefficients of the nine quark currents are different.

The modern theory of the weak interaction is based on the idea that, from the "correct point of view," the quark current must look quite similar to the lepton current. We shall use here a simplified example to elucidate what this "correct viewpoint" must be. Let us transfer to an imaginary world with only two generations: "electronic" (ν_e, e, u, d) and "muonic": (ν_μ, μ, c, s). Incidentally, this was the picture of the real world to many a physicist at the beginning of 1975 when the c quark was already discovered but the τ lepton and b quark were not.

The idea is to write the total current in this world in the form

$$j = \bar{e}\nu_e + \bar{\mu}\nu_\mu + \bar{d}'u + \bar{s}'c,$$

where d' and s' are "rotated quarks" formed as mutually orthogonal combinations

$$d' = d\cos\theta_c + s\sin\theta_c,$$
$$s' = -d\sin\theta_c + s\cos\theta_c.$$

As we shall see, the four coefficients $a_{\bar{d}u}$, $a_{\bar{s}u}$, $a_{\bar{d}c}$, $a_{\bar{s}c}$ are expressed through a single parameter: the angle θ_c, called the Cabibbo angle. It is remarkable that experimental data bear out this structure of the weak current (to within the corrections caused by the existence of the third generation). The experimental value of the angle θ_c is close to 13° ($|\sin\theta_c| \simeq 0.22$). In experiments, the smallness of the angle θ_c is manifested in the suppression of decays of strange particles (due to the current $\bar{u}s$) relative to the neutron β decay (of course, after normalization according to the law $G_F^2 \Delta^5$ to equal energy release) and the preference of charmed particles to decay into strange particles (due to the current $\bar{s}c$).

Obviously, if the angle θ_c were zero, strange particles would be absolutely stable, since an s quark could, in principle, decay only into a heavier c quark; in real decays, this is forbidden by the energy conservation law.

From the standpoint of weak currents, "true particles" are rotated states, d' and s', that have no definite masses. From the standpoint of mass, true particles are d and s, having definite and unequal masses. If d' and s' had definite masses, we would have to deal only with d' and s'. This situation would then resemble that of the neutrinos (see below).

Returning now to the real world with its three generations, we write

$$j = \bar{e}\nu_e + \bar{\mu}\nu_\mu + \bar{\tau}\nu_\tau + \bar{d}'u + \bar{s}'c + \bar{b}'t,$$

where the three primed quarks are rotated quarks, d,s,b→d',s',b', and, instead of a simple 2 × 2 rotation matrix, we have to operate with a 3 × 3 matrix whose general form is rather complicated. It can be shown that, in the general case, the elements of this matrix are expressed in terms of four independent parameters: three angles $\theta_1, \theta_2, \theta_3$ (Euler angles in three-dimensional space) and a phase factor $e^{i\delta}$.

The angle θ_1 is close to the Cabibbo angle (experimentally $|\sin\theta_1| \approx 0.231 \pm 0.003$), while the experimental values of the other two angles and the phase are still known rather poorly:

$$0.02 < |\sin\theta_2| < 0.09, \quad |\sin\theta_3| < 0.04, \quad |\delta| \leq 0.4.$$

It is very interesting and important to determine these parameters and test the whole scheme of rotated quarks. Later, we shall see that the idea of rotated quarks is important in the unified gauge theory of electromagnetic and weak interactions.

MIRROR ASYMMETRY

The three preceding sections were devoted to discussing what can be called the flavor structure of weak interactions and completely ignored the spatial and spin properties of charged currents. Now we shall take up those aspects.

The fundamental feature of the weak interaction is that weak processes are mirror-asymmetric.

In 1956 Lee and Yang, looking for a plausible interpretation of K^+ meson decays into two and three pions, which at the time seemed paradoxical, came up with the hypothesis that weak interactions do not conserve space reflection parity. Only a few months later, numerous experiments confirmed the correctness of this hypothesis. Mirror asymmetry was discovered in the β-decay of nuclei and in decays of muons, pions, K mesons, and hyperons. The striking fact is that it was not a tiny effect but a 100 percent asymmetry in dozens of different decays.

In hindsight, it is difficult to understand how this spectacular phenomenon remained unnoticed for such a long time, but it is easy to imagine the shock that this discovery created. Indeed, the parity conservation law was regarded as one of the great geometric laws of conservation, along with those of momentum and angular momentum. Conservation of momentum follows from the homogeneity of space, as conservation of angular momentum follows from its isotropy. Likewise, parity conservation seemed to be inevitable because of the presumably obvious reflection symmetry of empty space, that is, because the vacuum has no helical properties. But empty space proved far from simple and its properties far from obvious. The unusual properties of the vacuum in QCD have been discussed in earlier sections. Later we shall mention other examples.

It must be emphasized that the shock caused by parity nonconservation was largely of a philosophical nature. As for the tools and techniques available to the theorist, quantum field theory encompassed this phenomenon with no difficulties whatsoever.

The space parity, P, of a physical quantity characterizes the behavior of this quantity under a mirror reflection of coordinate axes, the so-called P-reflection: $x \to -x$, $y \to -y$, $z \to -z$. Vector quantities (so-called polar vectors), such as momentum **p**, vector potential **A**, and electric field strength **E**, reverse their sign under this transformation. They are P-odd. Pseudovector (or axial vector) quantities, such as the vector product of two vectors, for example orbital angular momentum **L**, spin **S**, and magnetic field strength **H**, do not change their sign. They are P-even.

The scalar product of two vectors or of two axial vectors is a scalar. Scalars are P-even. The scalar product of a polar and an axial vector is a pseudoscalar. Pseudoscalars are P-odd.

Until 1956, it was believed that the Lagrangian must be a scalar. After 1956, it became clear that the Langrangian of the weak interaction consists of two terms: a scalar and a pseudoscalar.

Because of the scalar term, a pseudoscalar K^+ meson decays into three pions, conserving parity. Because of the pseudoscalar term, the same K^+ meson decays into two pions, violating parity. In most other weak decays, both terms produce identical final particles but in different spin-orbital states. The above-mentioned reflection-asymmetric effects appear as a result of the interference of these states. Such are the longitudinal left-handed polarization of the spin of a β-electron, the correlation of the momentum of a β-electron with the spin of a decaying neutron, and others.

Such correlations are sign-reversing under P-reflection, so that we find certain mirror image processes absent in nature: for instance, the emission of a β-decay electron with positive (right-handed) longitudinal polarization.

$V-A$ CURRENT

In his original theory, Fermi postulated that weak currents are vector currents, that is, are four-dimensional vectors, like the electromagnetic current. Later, an erroneous conclusion was drawn from erroneous experimental data that weak currents are scalar and tensor currents. Only in

1957, after parity violation was discovered, did Feynman and Gell-Mann and, independently, Marshak and Sudarechan, and also Sakurai, suggest that each of the weak currents is the difference between a vector and an axial vector current. This current was named $V-A$ current (reads "vee minus ay").

The product of two $V-A$ currents naturally gives the sum of a scalar and a pseudoscalar in the weak interaction Lagrangian and, hence, explains parity violation.

As an example, consider the electron-neutrino current. Its vector component V is $\bar{e}\gamma_\alpha \nu_e$, its axial-vector component A is $-\bar{e}\gamma_\alpha\gamma_5\nu_e$ (the minus sign is in accord with a traditional definition), so that, in this case, the $V-A$ current has the form

$$\bar{e}\gamma_\alpha (1 + \gamma_5) \nu_e.$$

This is a suitable place to say a few words about the important role the matrix γ_5 plays in the theory of weak interactions. By definition, $\gamma_5 = i\gamma_0\gamma_1\gamma_2\gamma_3$, where $\gamma_0,\gamma_1,\gamma_2,\gamma_3$ are the four Dirac matrices. The quantity $\frac{1}{2}(1 + \gamma_5)$, applied to a four-component spinor ψ that describes a massless particle, selects its left-handed helicity component, ψ_L. The quantity $\frac{1}{2}(1 - \gamma_5)$ selects its right-handed component, ψ_R.

It is easy to show that $\bar{e}\gamma_\alpha (1 + \gamma_5) \nu_e = 2\bar{e}_L\gamma_\alpha\nu_L$. This means that all particles in the $V-A$ current, e, ν_e, μ, ν_μ, ..., u, d, ..., t, enter with their left-handed components: e_L, ν_{eL}, ..., u_L, d_L, ..., t_L, and all antiparticles enter with their right-handed components: $\bar{e}_R$, $\bar{\nu}_{eR}$, ..., $\bar{t}_R$. Usually, the $V-A$ current is said to be left-handed, which refers to the particles it involves (not antiparticles).

The statement that all charged currents necessarily have a $V-A$ structure was a daring one because, in 1957, it went against a number of experiments whose results seemed beyond doubt to most physicists, but, as it turned out, there was every reason for doubt. At present, not a single fact contradicts the universal $V-A$ structure of all charged currents.

We shall conclude this section by giving the expression for the Lagrangian for the charged current interaction

$$\mathscr{L}^{ch} = \frac{G_F}{\sqrt{2}} j_\alpha j_\alpha{}^+,$$

where $j_\alpha = 2\ (\bar{e}_L\gamma_\alpha\nu_{eL} + \bar{\mu}_L\gamma_\alpha\nu_{\mu L} + \bar{\tau}_L\gamma_\alpha\nu_{\tau L} + \bar{d}_L'\gamma_\alpha u_L + \bar{s}_L'\gamma_\alpha c_L + \bar{b}_L'\gamma_\alpha t_L)$ and $j_\alpha{}^+$ is the conjugate current.

C,P,T, SYMMETRIES

The P reflection is one of three closely interrelated discrete transformations. The other two transformations are time reversal, T, and charge conjugation, C. The invariance under time reversal $t \to -t$ demands that the probability amplitudes of the direct and reversed processes be equal. The invariance under charge conjugation demands that the amplitudes of two processes that differ only by the replacement of all particles by their antiparticles be equal.

Quantum field theory contains the fundamental Lüders-Pauli theorem, or CPT theorem, which states that it is impossible to construct a meaningful CPT-noninvariant Lagrangian. Hence, violation of P parity must be accompanied by violation of either C or T symmetry, or both.

Indeed, the very first experiments that revealed violation of mirror symmetry demonstrated that charge symmetry is also 100 percent violated in weak decays. Thus, decay electrons have predominantly left-handed polarization, while decay positrons in charge-conjugate decays have predominantly right-handed polarization.

The violation of both P and C symmetries is especially pronounced in the properties of massless neutrinos, which behave like ideal screws: All neutrinos have left-handed helicity and all antineutrinos have right-handed helicity.

The theory of longitudinally polarized fermions described by two-component spinors was first formulated by Weyl in 1929 and was immediately rejected because of its mirror asymmetry. It was revived by Landau, Salam, Lee, and Yang as the theory of two-component neutrinos in papers published at the beginning of 1957; this was an important landmark on the way to creating the $V-A$ theory. The $V-A$ theory described in the preceding section generalizes the theory of the Weyl neutrino to other fundamental fermions and contains the maximum possible violation of both the P and C symmetries.

A hope persisted for several years after the discovery of mirror and charge asymmetry that weak interactions would be merciful, at least to the CP symmetry and, by virtue of the CPT theorem, to the T reversibility. This hope stemmed from the fact that, within experimental accuracy, which was not better than several percent, the decays investigated were CP invariant. But, in 1964, Cronin, Christenson, Fitch, and Turlay discovered the decay of a long-lived neutral K meson into two π mesons: $K_L^\circ \to \pi^+\pi^-$. Since K_L° mesons mostly decay into CP-odd states of three pions and the

$\pi^+\pi^-$ state is CP even, the discovery of the $K_L^\circ \to \pi^+\pi^-$ decay meant that the CP symmetry was also violated.

A careful experimental and theoretical study of this and other decays of the K_L° meson (into $\pi^\circ\pi^\circ$, $e^\pm\nu\pi^\mp$, or $\mu^\pm\nu\pi^\mp$) confirmed the violation of the CP and T invariance but found no evidence of the violation of the CPT invariance.

In contrast to the P and C asymmetries, all known CP asymmetric effects are very small (roughly 10^{-3} in amplitude) and are restricted exclusively to the decays of K_L° mesons. As a result, the nature of the CP violation remains an unsolved problem.

An important step to finding the mechanism of CP violation would be the discovery of a nonzero electric dipole moment of the neutron, d_n, which is forbidden by T invariance (under T reversal the electric moment of a particle retains its sign, while its spin, to which this dipole moment is proportional, reverses its sign). The upper experimental bound reached at present for d_n is $d_n \lesssim e \cdot 4 \times 10^{-25}$cm, where e is the electron charge. Various mechanisms of CP violation offered by theorists predict the value of d_n to be in the range from $e \cdot 10^{-24}$ to $e \cdot 10^{-38}$cm.

We shall discuss here one of the possible mechanisms of CP violation which is quite popular with theorists. When mentioning "rotated quarks," we stressed that the coefficients of quark current depend on three Euler angles and a phase factor $e^{i\delta}$. It can be shown that the departure of the phase δ from zero (or π) signifies CP violation. Calculations show that the neutron dipole moment anticipated with this mechanism of CP violation is very small ($d_n \lesssim e \cdot 10^{-32}$cm), much too small for experimental detection. This model of CP violation can be verified by measuring to within a fraction of one percent the probabilities of the $K_L^\circ \to 2\pi^\circ$ and $K_L^\circ \to \pi^+\pi^-$ decays. Preparations for this very difficult experiment are now in progress.

To conclude the outline of CP symmetry, it must be mentioned that an interesting question is, whether or not CP symmetry is violated in quantum chromodynamics. The point is that not a single known general principle forbids us to add an additional term, which is usually written in the form

$$\frac{\theta \alpha_s}{16\pi} F_{\alpha\beta}^a F_{\gamma\delta}^a \epsilon_{\alpha\beta\gamma\delta}$$

which is *CP*-odd, to the standard QCD Lagrangian. Here $F_{\alpha\beta}^a$ is the tensor of gluon field strength ($a = 1,2,\ldots 8$), $\epsilon_{\alpha\beta\gamma\delta}$ is an antisymmetric tensor, θ is a dimensionless coefficient (sometimes it is called the vacuum angle), and α_s is the familiar constant of strong coupling. This term, usually referred to as the θ-term, is C-even, P-odd, and, hence, CP-odd (it is similar to the scalar product of the electric and magnetic fields, **EH**). It can be shown that the available experimental upper bound on the neutron dipole moment implies $\theta < 10^{-8}$. It is extremely interesting to try to explain whey the θ-term is so small. Among suggested explanations, a very light neutral pseudoscalar particle, the axion, was invented to account for the smallness of θ. An experimental search for the axion has not confirmed the existence of this particle.

NEUTRAL CURRENTS

The foregoing discussion of the weak interaction dealt with processes produced by charged currents.

In 1973, muonless neutrino reactions due to the coupling of the so-called neutral currents were discovered in neutrino reactions. In these reactions, muon neutrinos that collided with nucleons and transferred to them part of their energy did not convert into muons but presumably remained muon neutrinos, as, for example, in the reaction

$$\nu_\mu + p \rightarrow \nu_\mu + p + \pi^+ + \pi^-.$$

The observation of these reactions has led to the conclusion that there is a coupling of the neutral neutrino current $\bar{\nu}_\mu \nu_\mu$ to neutral quark currents of the type $\bar{u}u$ and $\bar{d}d$. The coupling constant of this interaction was found to be approximately the same as for charged currents, that is, G_F.

The search for other neutral currents led to the discovery in 1978 of the electron current $\bar{e}e$. This weak P-odd current was first detected in the rotation of the polarization plane of a laser beam passing through atomic bismuth vapor in Barkov and Zolotorev's experiment at Novosibirsk. The optical activity of bismuth vapor signifies that there is a weak parity-violating interaction between atomic electrons and nuclei, that is, between electrons and u and d quarks. Somewhat later, the coupling of the ēe current to the ūu and d̄d currents was observed in the scattering of longitudinally polarized electrons on deuterons at the Stanford Liner Accelerator.

Finally, in 1982, the coupling between the ēe current and the $\bar{\mu}\mu$ and $\bar{\tau}\tau$ currents was discovered. I mean here the observation of the weak charge asymmetry in the reactions $e^+e^- \rightarrow \mu^+\mu^-$ and $e^+e^- \rightarrow \tau^+\tau^-$ at the PETRA collider. All neutral currents found so far conserve the flavor of the participating particles; they are diagonal, that is, they transform the particle into itself: electron into electron, muon into muon, and so on. Flavor-changing currents of the type ēμ or d̄s have not been found. Later, we shall see that this is in agreement with the theory that predicts the existence of twelve diagonal currents,

$\bar{e}e$, $\bar{\mu}\mu$, $\bar{\tau}\tau$, $\bar{\nu}_e\nu_e$, $\bar{\nu}_\mu\nu_\mu$, $\bar{\nu}_\tau\nu_\tau$, $\bar{u}u$, $\bar{d}d$, $\bar{s}s$, $\bar{c}c$, $\bar{b}b$, $\bar{t}t$.

The helicity structure of neutral currents is more complicated than that of charged currents. Experiments show that the total neutral current j_α^n consists of terms of two types: left-handed $\bar{\psi}_L\gamma_\alpha\psi_L$ and right-handed $\bar{\psi}_R\gamma_\alpha\psi_R$.

The left-handed currents of the upper particles, ν_e, ν_μ, ν_τ, u, c, t, enter with the coefficient ($+\frac{1}{2} - Q\sin^2\theta_W$), where Q is the particle charge and θ_W is the so-called Weinberg angle (see below). The left-handed currents of the lower particles, e, μ, τ, d, s, b, enter with the coefficient ($-\frac{1}{2} - Q\sin^2\theta_W$). The coefficients with the right-handed currents are identical for the upper and lower particles and equal $-Q\sin^2\theta_W$.

In the next chapter, we show that this structure of the current implies that left-handed particles form doublets with respect to the weak isospin group, (ν_{eL}, e_L), (u_L, d_L), etc., while the right-handed particles, ν_R, e_R, u_R, d_R, etc., are isotopic singlets. (To avoid possible misunderstandings, it must be emphasized that the weak isospin we discuss here is in no way related to the ordinary isospin of hadrons that we considered in Chapter III.)

The experimental value of the angle θ_W is $\sin^2\theta_W \approx 0.22$ (a mnemonic relation is $\sin^2\theta_W \approx \sin\theta_c \approx 0.22$; note that $\theta_W \neq \theta_c$). The Lagrangian for neutral currents interaction is

$$\mathscr{L}^n = \frac{G_F}{\sqrt{2}} j^n j^n.$$

Neutral currents were predicted by the unified theory of electromagnetic and weak interactions. The experimental discovery of neutral currents marked the triumph of this theory. The above-described structure of the currents will be better understood after we have outlined the foundations of the electroweak theory in the next chapter.

First, however, we shall turn to neutrino masses. Much work has been done in this field in recent years. The problem is essentially similar to that of rotated quarks; thus, it would be more logical to discuss it immediately after the section on components of charged current. I chose to place it at the end of this chapter because, in contrast to the subjects of the preceding sections, the experimental situation remains quite uncertain in this case.

NEUTRINO MASSES AND OSCILLATIONS. DOUBLE β-DECAY

When comparing the lepton and quark currents, we emphasized that the former are much simpler and that they owe their simplicity to the zero mass of the neutrino. However, there is a justified suspicion that this simplicity is illusory: that, in reality, neutrino masses are nonzero, that the so-called neutrino oscillations, that is, transitions between distinct types of neutrino, take place in vacuum and, furthermore, that the neutrino and antineutrino are not separated by such a clearcut boundary.

Until very recently, laboratory experiments gave no indication that neutrino masses are nonzero, but high accuracy was achieved only for the electron neutrino ($m_{\nu_e} < 35$ eV). The upper bounds for the muon neutrino and, especially, for the τ neutrino are much higher: $m_{\nu_\mu} \lesssim 0.6$ MeV and $m_{\nu_\tau} \lesssim 250$ MeV. There is, however, a cosmological upper bound on the masses of all sorts of neutrino, according to which the sum $m_{\nu_e} + m_{\nu_\mu} + m_{\nu_\tau}$ is definitely below 100 eV.

This restriction is obtained by the following argument, suggested by Gershtein and Zeldovich. According to the Big Bang theory, the number of primordial neutrinos must be approximately equal to the number of primordial photons. These latter were discovered in the form of background radiation in 1965 by Penzias and Wilson. The universe contains about 10^9–10^{10} photons per proton. If there are as many neutrinos as photons and the mass of each neutrino is, say, 100 eV, it is evident that the total mass of the neutrino gas in the universe should exceed the mass of ordinary matter by two to three orders of magnitude. Calculations show that such a high density should have resulted in a faster evolution of the universe, and the age of the universe thus obtained would be shorter than the age of certain terrestrial rocks. But it must be realized that the cosmological restriction on the masses of ν_μ and ν_τ is invalidated if these particles decay sufficiently rapidly into $\nu_e + \gamma$.

In 1980, results were published of the experiment carried out at the ITEP in Moscow, according to which the electron neutrino mass is distinct from zero (14eV $\leq m_{\nu_e} \leq$ 46eV). This conclusion was drawn from measuring the electron spectrum in tritium β-decay, ^{3}H$\rightarrow ^3$He $+ e^- + \tilde{\nu}_e$.

When an electron is ejected with nearly maximum energy, the kinetic energy of a neutrino is nearly zero. This creates optimal conditions for detecting the possible neutrino mass. By measuring the shape of the electron spectrum close to its upper bound, experimenters obtained the result given above. Tritium decay, with its uniquely small energy release, is especially suitable for such measurements.

Unfortunately, a problem cannot be considered solved by a single experiment. This is especially true when the experiment reaches the limiting accuracy possible with the present state-of-the-art of experimental techniques, as in the ITEP experiment. New independent experiments are needed. Experiments designed to establish the neutrino mass are being prepared and run now in a number of laboratories. We can anticipate that, in the next few years, the problem of whether the electron neutrino mass is in the range 10–30 eV will be finally solved.

Astrophysicists were probably pleased more than anybody else with the publication of the news that the neutrino mass may be nonzero. Massive neutrinos are needed, they say, for at least two different reasons. First, to explain the nature of invisible massive coronas in galaxies and galaxy clusters. The existence of invisible mass in and around galaxies has been actively discussed by astronomers for over a decade. Clouds of massive neutrinos fit very well into this picture. Second, certain difficulties in the theory of the formation of galaxies proved to be resolvable by the same neutrino clouds. In both cases, neutrinos with a mass of about 10–30 eV are most suitable.

Obviously, to conclude that the electron neutrino mass indeed lies in this range on the basis of the above-given astrophysical arguments would be more than premature. Perhaps the needs of astrophysics could be met not only by neutrinos, but also by any neutral particles with masses from 10 to 30 eV (e.g., photinos or gravitinos, which will be discussed in the section devoted to supersymmetry).

From the purely theoretical point of view, we do not see at present any reason to assume zero neutrino masses. In this respect, the views reigning in the theoretical physics community have changed radically over the past decade. Formerly, it seemed more natural to expect that the neutrino mass was zero rather than some small number, because "all coefficients in physics are of the order of unity, so why should we expect a small parameter here?" At present, the widely popular point of view is that the existence of a massless particle calls for strict local symmetry. And since there is no such symmetry in the case of the neutrino, its mass can hardly be zero.

As for the value of the expected neutrino mass, the situation is much less definite. There is no unanimity among theorists, but the majority, relying on grand unification models, consider it more natural for the neutrino mass to be very small, say, around 10^{-5} eV. This number is obtained by dividing the square of the τ lepton mass (~ 1 GeV^2) by the grand unification mass ($\sim 10^{14}$ GeV; more on grand unification later). If the electron is taken instead of the τ lepton, the number will be less by seven orders of magnitude. Thus, these guesses cannot be taken too seriously.

In recent years, the neutrino mass problem generated much interest in the search for two new phenomena: neutrino oscillations and double β decay.

The possibility of neutrino oscillations was pointed out for the first time by Pontecorvo in the middle of the 1950s, soon after Pais and Piccioni predicted oscillation effects in beams of neutral K mesons. At the present moment, the number of papers devoted to a theoretical discussion of neutrino oscillations runs into the hundreds. The experimental search for this phenomenon is being carried out in a number of nuclear reactor and accelerator laboratories.

Here is a brief explanation of the principle of oscillations with a simplified example of two neutrinos, ν_e and ν_μ. Assume that the states ν_e and ν_μ entering weak currents do not have definite masses but represent orthonormalized quantum-mechanical superpositions of two other states, ν_1 and ν_2, with definite masses, m_1 and m_2:

$$\nu_e = \nu_1 \cos\alpha + \nu_2 \sin\alpha,$$
$$\nu_\mu = -\nu_1 \sin\alpha + \nu_2 \cos\alpha.$$

Here the angle α is similar to the Cabibbo angle in the case of the d and s quarks.

As an example, consider a beam of ν_μ. Let the beam have a definite momentum **p**. By virtue of the mass difference, the energies of ν_1 and ν_2 will be different:

$$E_1 - E_2 = \sqrt{\mathbf{p}^2 + m_1^2} - \sqrt{\mathbf{p}^2 + m_2^2} \approx \frac{m_1^2 - m_2^2}{2E}.$$

As a result, the relative phase of ν_1 and ν_2 increases linearly with time and an admixture of ν_e will gradually appear in the originally pure beam of ν_μ. We readily calculate that the fraction of this admixture is a periodic function of distance,

$$\sin^2 2\alpha \sin^2 \left(1.27 \frac{\delta m^2 L}{E}\right),$$

where E is the neutrino energy (in MeV), L is distance between the neutrino source and the detector (in meters), and $\delta m^2 = m_1^2 - m_2^2$ (in eV^2).

If beam neutrinos have sufficiently high energy, as they typically have in accelerator experiments, oscillations can be detected by observing two effects in the beam-target interaction. First, by the appearance of neutrinos of a new kind, and second, by a decrease in the number of original neutrinos. In the case of reactor (anti)neutrinos, $\bar{\nu}_e$, the first effect is unobservable because the energy of reactor antineutrinos is below the threshold of the reaction $\bar{\nu}_\mu + p \rightarrow \mu^+ + n$. The search for the second effect, the effect of the leakage of the initial (anti)neutrinos, may prove successful.

Observations of such leakage were reported in 1980 by a group at the Savannah River reactor in the USA. However, subsequent measurements with reactor neutrinos in France and Switzerland did not confirm the existence of the effect and yielded the restrictions $\delta m^2 \lesssim 10^{-2} eV^2$ for $\sin^2 2\alpha \sim 1$, $\delta m^2 \lesssim 1$ eV^2 for $\sin^2 2\alpha \sim 0.1$.

So far, attempts to observe oscillations with accelerators also failed to yield a positive result. No oscillations were observed with neutrinos created by cosmic rays in the earth's atmosphere. The most accurate measurements of this type were realized at the Baksan neutrino observatory. Reactions observed there were initiated by neutrinos that had been created over Australia and had crossed the globe. Despite the long path from the source to the detector, no signs of leakage, in comparison with the calculated neutrino flux, were found.

The search for neutrino oscillations is still going on. Of course, when and if oscillations are finally discovered, their study will not be limited to the system ν_e and ν_μ, but will also cover ν_τ. Like the description of three quark currents, the description of three lepton currents will require a 3×3 matrix that depends on three Euler angles and a phase. One has to keep in mind that a description of lepton currents may prove to be even more complicated than that of quark currents. The possibility of complications may appear because, in contrast to quarks, neutrinos are electrically neutral.

In the case of quarks, the mass terms in the Lagrangian can be only of the type $m\bar{\psi}\psi$, transforming a particle into another particle. This is what we call the Dirac mass. But in the case of the neutrino, particles may have, along with the Dirac masses, the so-called Majorana masses $m'\psi C \psi$ (where C is the charge conjugation matrix), transforming a particle into an antiparticle (this term is forbidden for quarks because the quark and antiquark charges are different).

Leptons are usually characterized by the lepton quantum number L, equal to +1 for e^-, μ^-, τ^-, ν_e, ν_μ, and ν_τ, and −1 for e^+, μ^+, τ^+, $\tilde{\nu}_e$, $\tilde{\nu}_\mu$, and $\tilde{\nu}_\tau$. Lepton number is conserved in the standard theory of the weak interaction. If, however, the neutrinos possess Majorana masses, the lepton number is not conserved. Besides, instead of three neutrinos and antineutrinos we would have to deal with six truly neutral, the so-called Majorana, neutrinos. The neutral states entering weak currents would be superpositions of these Majorana neutrinos.

Nonconservation of the lepton number makes possible a very peculiar effect: neutrinoless double β decay. In ordinary β decay, one d quark undergoes a weak transition into a u quark. In double β decay, two d quarks transform simultaneously into two u quarks. Two situations are possible: the two-neutrino decay 2β (2ν), when two (anti)neutrinos are

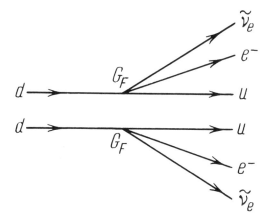

FIGURE 27. The mechanism of the double β decay with the emission of two (anti)neutrinos.

emitted (see Figure 27), and neutrinoless decay 2β (0ν), when a virtual neutrino emitted by one quark is absorbed by another quark (see Figure 28). This last process is possible only with the Majorana neutrino, because the lepton charge is not conserved in this process. Both these decays occur in the second order of perturbation theory in the weak coupling constant G_F; thus, the expected half-lives, $T_{\frac{1}{2}}$, are very large.

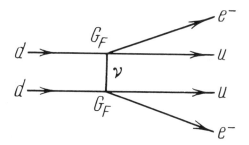

FIGURE 28. The mechanism of the neutrinoless double β decay.

The two-neutrino decay rate can be calculated more or less reliably (it varies considerably with the kind of nucleus because of high sensitivity to the nuclear structure and to energy release). In contrast to this, the neutrinoless decay rate cannot be predicted as long as we do not know the degree and the mechanism of lepton number nonconservation (it can be shown that the neutrinoless decay rate must be proportional either to the square of the Majorana mass of the neutrino or to the squared coupling constant of the hypothetical charged right-handed currents).

Neither neutrinoless nor two-neutrino decays have been confidently observed in experiments. Actually, certain data were published in 1980 which their authors interpret as a possible indication of the 2β (2ν) decay $^{82}\text{Se}_{34} \rightarrow {}^{82}\text{Kr}_{36}$ with a half-life $T_{\frac{1}{2}}^{82} = 10^{19}\text{--}10^{20}$ years. However, other data give $T_{\frac{1}{2}}^{82} > 10^{21.5}$ years. Lower bounds for laboratory data on $T_{\frac{1}{2}}^{48}$ ($^{48}\text{Ca}_{20} \rightarrow {}^{48}\text{Ti}_{22}$) and $T_{\frac{1}{2}}^{76}$ ($^{76}\text{Ge}_{32} \rightarrow {}^{76}\text{Se}_{34}$) are approximately of the same order of magnitude. These inequalities can be used to establish upper bounds on the neutrino's Majorana mass. These bounds are different with different authors and vary from tens to hundreds of electron volts.

It appears promising to look for the double β decay in other nuclei, for instance, $^{136}\text{Xe}_{54}$, $^{100}\text{Mo}_{42}$, and $^{116}\text{Cd}_{48}$, and also to search for electron capture from the atomic K shell, accompanied by the emission of a positron,

$$e^- + (A, Z+2) \rightarrow e^+ + (A, Z)$$

(in transitions of the type

$$^{99}\text{Ru}_{44} \rightarrow {}^{99}\text{Mo}_{42}, \quad {}^{106}(\text{Cd}_{48} \rightarrow {}^{106}\text{Pd}_{46}, \quad {}^{124}\text{Xe}_{54} \rightarrow {}^{124}\text{Te}_{52},$$
$$^{130}\text{Ba}_{56} \rightarrow {}^{130}\text{Xe}_{54}, \quad {}^{138}\text{Ce}_{58} \rightarrow {}^{138}\text{Ba}_{56}).$$

There are indirect geochemical data on the concentration of $^{130}\text{Xe}_{54}$ and $^{128}\text{Xe}_{54}$ isotopes in natural tellurium that are interpreted as an indication that the $^{130}\text{Te}_{52} \rightarrow {}^{130}\text{Xe}_{54}$ decays (with $T_{\frac{1}{2}}^{130} \approx 10^{21.3}$ years) and $^{128}\text{Te}_{52} \rightarrow {}^{128}\text{Xe}_{54}$ decays occur and that $T_{\frac{1}{2}}^{130} : T_{\frac{1}{2}}^{128} \approx 6.1 \times 10^{-4}$ (the expected value of this ratio for the neutrinoless decay induced by the Majorana mass is approximately 1.25×10^{-2}, and approximately 1.5×10^{-4} for the two-neutrino decay). I am of the opinion that it would be premature to draw from these data any conclusions on the neutrino mass. But some authors conclude that $m' = 10\text{--}30$ eV.†

The reader may have formed an impression that we are paying too much attention to particular details in this chapter, while, say, in the preceding chapter devoted to strong interactions, many problems of the same rank were mentioned only in passing. However, this impression is not quite correct. In both chapters, we concentrate on the structure of the relevant Lagrangians.

---

†*Note added after the book was completed.* New measurements (Kirsten et al., 1982) do not confirm earlier results and give $T_{\frac{1}{2}}^{130} / T_{\frac{1}{2}}^{128} = (0.90 \pm 0.95) \times 10^{-4}$. New calculations of the expected value of this ratio give 4.4×10^{-2} in the case of 2β (0ν) and 3.3×10^{-4} in the case of 2β (2ν).

ON THE RELIABILITY OF EXPERIMENTAL DATA

The physics of elementary particles is done by people. It is characteristic of man to err: mistakes are made both by experimenters and by theorists. Examples of such mistakes were mentioned earlier in the book. Thus, the current structure of the β-decay interaction was initially determined incorrectly and erroneous figures for the probabilities of the decays $\pi \rightarrow e\nu$, $K° \rightarrow \mu^+\mu^-$, and $K° \rightarrow 2\pi°$ were regarded as correct for a long time. More recent examples could also be given.

Why then do physicists regard a multitude of phenomena as experimentally established, despite such mistakes? Could similar mistakes crop up in the experiments on which we unreservedly rely today? How can it be guaranteed that these experiments are correct if so many incorrect results occurred in the past?

The only guarantee is to accept a result as reliable only if it is obtained independently by several different groups employing different experimental methods. This condition is absolutely necessary but may not be sufficient, and does not provide a 100 percent guarantee. The 100 percent guarantee appears when the phenomenon recedes from the frontline of the science, when it is reproduced routinely, with the statistics of events exceeding by thousands or millions that with which the discovery was made, and when the quantities characterizing the phenomenon become known to an accuracy of several decimal places. Another way is not so much quantitative as it is qualitative: the search and discovery of a number of related phenomena that often follow the original discovery. Both these paths are well traced in the discoveries of P violation, CP violation, charm, and so on.

One of the difficulties of work in the field of high energy physics is the fact that very preliminary (not necessarily correct) results often provoke a premature general discussion. The guilt partially lies with theoretical physicists who snatch the hot experimental data "right off the griddle." This frequently leads to an expenditure of very considerable efforts to give an explanation of a "result" that in a year or two bursts like a soap bubble. Of course, competition betweeen experimental groups also plays an important role. But, somehow or other, in several years the truth usually comes to the surface and turbulence evolves to a quiet clarity.

Chapter 5

The Electroweak Theory

Distinctive features of the weak interaction
SU(2) × U(1) symmetry
The photon and the Z boson
Coupling of charged currents
Coupling of neutral currents
The search for the W and Z bosons
Symmetry breaking
Higgs bosons
Models, models . . .
Scalars: Problem no. 1
On the development of the theory

DISTINCTIVE FEATURES OF THE WEAK INTERACTION

The distinctive features of weak processes discussed in the preceding chapter are the following.

1. Their weakness (slow rates), manifested mn the fact that the rates of these processes are many orders of magnitude less than the rates of the strong and electromagnetic processes.

2. Their small interaction radius, at least two orders of magnitude smaller than the strong interaction radius. No departures from the exact four-fermion interaction have been detected so far (up to January 1983) in any of the weak processes.

3. The strong, maximum possible, violation of space reflection and charge conjugation parities. Thus, charged currents include only the left-handed components of spinors that describe the particles, ψ_L, and only the right-handed components of spinors that describe the anti-particles, $\bar{\psi}_R$.

4. CP violation.
5. Nonconservation of flavors (strangeness, charm, etc.).
6. The fact that neutrinos participate only in weak interactions.

It is all the more astonishing, then, that, despite such striking differences, the weak and electromagnetic interactions appear to be manifestations of one and the same interaction, which has recently been given the name "electroweak interaction."

According to electroweak theory, the weak charged currents interact by the exchange of W bosons and the neutral currents interact by the exchange of Z bosons, just as electromagnetic currents interact by the exchange of photons. The weak interaction is weak and its radius is small because, in contrast to photons, the W and Z bosons are very heavy particles. The other features of the weak interaction are directly contained in the assumption of the form of the original fermion currents of the electroweak theory. We must therefore be surprised not with the mirror and charge asymmetry of the weak interaction in the electroweak theory, but with the mirror and charge symmetry of the electromagnetic interaction.

SU(2) × U(1) SYMMETRY

The theory of the electroweak interaction was constructed in the 1960s. In 1979, Glashow, Salam, and Weinberg were awarded the Nobel prize for their contributions to this achievement. The foundation of the theory is the gauge symmetry SU(2) × U(1). Here SU(2) is the group of weak isospin and U(1) is the group of weak hypercharge. The electroweak theory differs from pure electrodynamics and from chromodynamics in two characteristic features.

First, the gauge symmetry SU(2) × U(1) is spontaneously broken; consequently, the weak gauge bosons, the so-called intermediate bosons $W^{\pm}$ and Z°, are massive.

Second, the theory is mirror-asymmetric from the very beginning. This asymmetry lies in the foundations of the theory: The left-handed components of the fermions $\psi_L = \frac{1}{2}(1 + \gamma_5)\psi$ form isotopic doublets under the group SU(2),

$$\begin{pmatrix}u\\d'\end{pmatrix}_L, \begin{pmatrix}c\\s'\end{pmatrix}_L, \begin{pmatrix}t\\b'\end{pmatrix}_L, \begin{pmatrix}\nu_e\\e\end{pmatrix}_L, \begin{pmatrix}\nu_\mu\\\mu\end{pmatrix}_L, \begin{pmatrix}\nu_\tau\\\tau\end{pmatrix}_L.$$

At the same time, the right-handed components $\psi_R = \frac{1}{2}(1-\gamma_5)\psi$ of the same 12 fermions are isotopic singlets (primes mark the rotated quarks discussed in Chapter IV). Note that the weak isospin has no relation to the global isotopic symmetry that characterizes strong interactions. The same is true for the weak hypercharge.

The unbroken local symmetry SU(2) × U(1) calls for the existence of four massless vector bosons: two charged bosons, W^+ and W^-, and two neutral onces, $W°$ and $B°$. The three W bosons form a triplet and the $B°$ boson a singlet under the group SU(2). The W bosons are gauge fields of the weak isospin group SU(2). Their interaction is characterized by a weak "charge," namely, the gauge coupling constant g_2. The $B°$ boson constitutes the gauge field of the weak hypercharge group U(1). Its coupling is characterized by another "charge", g_1.

THE PHOTON AND THE Z BOSON

From the standpoint of the unbroken group structure SU(2) × U(1), the field of the photon, A, and that of the Z boson, Z, are less fundamental than the fields $W°$ and $B°$, being orthonormal linear superpositions of the latter,

$$A = B° \cos \theta_W + W° \sin \theta_W,$$
$$Z = -B° \sin \theta_W + W° \cos \theta_W,$$

where θ_W is the Weinberg angle. We shall soon see that $\tan \theta_W = g_1/g_2$. The superpositions A and Z are singled out in that one of them, A, remains massless under spontaneous symmetry breaking, while the other, Z, becomes massive, just as do the other two "fallen angels," W^+ and W^-.

This form of the fields A and Z can be found readily as soon as one takes into account the fact that spontaneous breaking of the symmetry SU(2) × U(1) does not affect the symmetry U(1)$_{em}$ responsible for the conservation of electric charge Q. Let us show this.

We begin with the covariant derivative D_μ. Taking into account the facts that the source of the triplet of fields $W = W^+, W^-, W°$ is the isospin **T** and that the source of the field $B°$ is the hypercharge Y (or, rather, $\frac{1}{2}Y$), we arrive at the following expression for the covariant derivative:

$$D_\mu = \partial_\mu - i(g_1 \tfrac{1}{2} Y B°_\mu + g_2 \mathbf{T} W_\mu)$$

Now we take into account the fact that, by definition, the charge Q (in units of e), the hypercharge Y, and the third projection of isospin T_3 are related by the formula

$$Q = T_3 + Y/2$$

and single out the field A, whose source is the electric charge Q and the field Z orthogonal to A. As a result, the amplitude of the emission of the fields γ, Z, W^+, and W^- by the corresponding sources is given by the expression

$$i \frac{g_1 g_2}{\sqrt{g_1^2 + g_2^2}} QA - i\sqrt{g_1^2 + g_2^2} (T_3 - Q \sin^2 \theta_w) Z$$
$$- ig_2 (T^- W^+ + T^+ W^-).$$

Simple arithmetic confirms that A and Z must indeed be described by the foregoing superpositions of the fields B° and W°. This expression is the central formula of the electroweak theory: it contains all information on the electromagnetic interaction (first term), neutral currents (second term), and charged currents (third term). It also implies that the electromagnetic coupling constant e is expressed through the constants g_1 and g_2+

$$e = \frac{g_1 g_2}{\sqrt{g_1^2 + g_2^2}} = g_2 \sin \theta_w.$$

COUPLING OF CHARGED CURRENTS

The constant g_2 characterizes the emission and absorption of $W^\pm$ bosons as e characterizes the emission and absorption of photons. Clearly, $g_2 > e$; hence, the weak interaction is, in fact, stronger than the electromagnetic. The fact that the weak interaction in the processes observed up to now was many orders of magnitude weaker than the electromagnetic interaction is explained by the very large masses of the W bosons.

As an example, consider the coupling of the currents $\bar{e}_L \gamma_\alpha \nu_{eL}$ and $\bar{\nu}_{\mu L} \gamma_\alpha \mu_L$ that produce the muon decay. Figure 29 shows how this interaction is realized via the exchange of a W boson. In Figure 30, the same

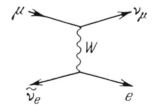

FIGURE 29. The decay of the muon through a virtual W boson.

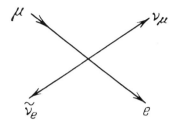

FIGURE 30. The decay of the muon in the approximation of a local four-fermion interaction.

interaction is shown as a local coupling of two currents in a single point. This approximation is good when the square of the 4-momentum transfer by the W boson, q^2, is small compared to the square of its mass m^2_W. In this case, the propagator of the W boson becomes independent of the momentum transfer,

$$\frac{1}{m^2_W - q^2} \to \frac{1}{m^2_W}.$$

By using a spatial description, we can say that, in this case, a heavy virtual boson goes from the point of emission to the point of absorption across a distance negligibly small compared to the wavelengths of the lighter particles taking part in the process.

By looking at Figures 29 and 30 we can easily express G_F through g_2 and m_W,

$$G_F = \frac{g_2^2}{4\sqrt{2}\, m^2_W} = \frac{\pi\alpha}{\sqrt{2}\, m^2_W \sin^2\theta_W},$$

where $G_F = 1.17 \times 10^{-5}$ GeV^{-2} is the Fermi constant, and $\alpha = e^2/4\pi = 1/137$. We have mentioned in the preceding chapter (and shall discuss in the next section) that the quantity $\sin^2\theta_W$ can be found from experiments with neutral currents. These experiments yield $\sin^2\theta_W \approx 0.22$. By inverting the formula for G_F, we can thus predict the mass of the W bosons,

$$m_W = \left(\frac{\pi\alpha}{\sqrt{2}\, G}\right)^{1/2} \frac{1}{\sin\theta_W} = \frac{37.3}{\sin\theta_W} \text{ GeV} \approx 80 \text{ GeV}.$$

It is of principle importance here that $\sin\theta_W$ determines not only the relation between g_2 and e but also the form of the neutral currents.

COUPLING OF NEUTRAL CURRENTS

Let us return to the central formula (*see* p. 66) and consider the term with the Z boson. It is readily verified that this term yields the form of the current described at the end of the section "Neutral Currents" in the previous chapter. Indeed, for all left-handed components, $T_3 = +1/2$ for the upper particles, and $T_3 = -1/2$ for the lower particles. Consequently, left-handed currents for upper and lower particles have the form

$$(\tfrac{1}{2} - Q \sin^2\theta_W)\bar\psi_L \gamma_\alpha \psi_L \text{ and } (-\tfrac{1}{2} - Q \sin^2\theta_W)\bar\psi_L \gamma_\alpha \psi_L,$$

respectively. The isotopic spin of "right-handed particles" is zero; thus, the central formula gives the expression

$$-Q \sin^2\theta_W \bar\psi_R \gamma_\alpha \psi_R$$

The weak interaction of neutral currents is realized through the exchange of virtual Z bosons. Note that the constant of emission of a Z boson, $\sqrt{g_1^2 + g_2^2}$, is greater than the constant of emission of a W boson, g_2. Their ratio is $1/\cos\theta_W$. However, the same central formula tells us that the ratio of the masses of the Z and W bosons is also equal to $1/\cos\theta_W$. Consequently, the effective four-fermion constant for neutral and charged currents is the same

$$G_F = \frac{g_2^2}{4\sqrt{2}\, m_W^2} = \frac{g_1^2 + g_2^2}{4\sqrt{2}\, m_Z^2}.$$

Let us mention how flavors are involved. Since weak isotopic fermion doublets contain rotated quarks, d', s', and b', charged currents contain transitions between quarks of different generations. This is not the case for neutral currents: the electroweak theory has no flavor-changing neutral currents. The reason for this is that weak currents have identical forms for all lower quarks and thus enter the total neutral current as a sum $\bar d' d' + \bar s' s' + \bar b' b'$. It is not difficult to show, taking the unitarity of the matrix relating the primed and nonprimed quarks into account, that

$$\bar d' d' + \bar s' s' + \bar b' b' = \bar d d + \bar s s + \bar b b;$$

this means that neutral currents are diagonal in flavors. The same is obviously true for the electromagnetic current.

THE SEARCH FOR THE W AND Z BOSONS

The discovery of the W and Z bosons would be a decisive step in the verification of the unified theory of the weak and electromagnetic interactions. The masses and partial widths of the decay channels of these particles, as well as their production cross-sections, are completely predicted by the theory.

A special proton-antiproton collider was built at CERN to produce W and Z bosons and began operations in the summer of 1981. The energy of each of the colliding beams is 270 GeV. This is quite sufficient to create W and Z bosons, whose expected masses are close to 80 and 90 GeV, respectively.

In the process of the production of, say, the W^+ boson, a u quark of a proton collides with a $\tilde{d}$ quark of an antiproton: $u+\tilde{d} \rightarrow W^+$. A monochromatic beam of protons can be regarded as a beam of quarks with a wide distribution of momenta. An antiproton can be viewed similarly. In the process of creating a W boson, a quark "chooses" an antiquark with a suitable momentum. The production of W bosons is best observed in the leptonic decays

$$W^{\pm} \rightarrow e^{\pm}\nu_e \text{ or } W^{\pm} \rightarrow \mu^{\pm}\nu_{\mu},$$

which yield single charged leptons with high transverse momenta.

The projected luminosity of the CERN collider is $\sim 10^{30}\,\text{cm}^{-2}\text{s}^{-1}$. Unfortunately, the actual luminosity over the first year of operations was around $10^{26}\,\text{cm}^{-2}\text{s}^{-1}$, not enough to observe the rare events of the creation of W and Z bosons. The collider was expected to reach the luminosity of $10^{28}\,\text{cm}^{-2}\text{s}^{-1}$, by the end of 1982. At that luminosity, it should be possible to observe one leptonic decay of the W boson per day and one leptonic decay of the Z boson per 10 days.†

A still bigger proton-antiproton collider, called TEVATRON, is being built at the Fermi National Laboratory (Fermilab) in Batavia, Ill. The energy of each beam will be 1 TeV$=10^3$ GeV. An acceleration-storage complex (UNK) near Serpukhov, with an energy of each of the colliding beams up to 3 TeV, is planned for completion in the 1990s.

Great expectations are connected with the LEP electron-positron collider, construction of which began at CERN in 1983 and which is expected to be operational at the end of 1988. Here it will be possible to observe the resonant formation of Z bosons: $e^+e^- \rightarrow Z^0$, with the energy of each of the colliding beams equal to $m_z/2$. It is estimated that LEP will produce one Z boson every 2–3 seconds. This will be a real factory of Z bosons.

A plan to develop an electron–positron collider for producing Z bosons was suggested at the Stanford Linear Accelerator Center (SLAC). Here it was suggested that the so-called Single Path Collider, also known as the

---

†See Author's note to p. 113.

Stanford Linear Collider (SLC) be built on the foundations of the existing linear accelerator.

One of the foremost problems in the study of Z bosons is to measure the total width of these particles. If, in addition to three known generations of fermions, there exist other generations, then Z bosons must be coupled to these "generations to come" in the same way as to those already known. And even if the masses of heavy charged leptons and quarks are so large that Z bosons cannot decay into pairs of these particles, the decay into pairs of "neutrinos to come" is unavoidable. Hence, the total width of the Z boson will serve as a "counter" of the total number of neutrino species.

The energy of LEP beams will presumably be raised in the future, and still more energetic (though not circular, but linear) colliders of electron-positron beams will be constructed, making possible the study of subtler but very important predictions of the electroweak theory dealing with the production of pairs of W^+W^- bosons.

SYMMETRY BREAKING

We have entered the edifice of the electroweak theory through the main portal marked "Gauge Symmetry," where everything looks beautiful. But the edifice has another entrance, not less important, but which at present looks much less beautiful.

In fact, the strict gauge symmetry $SU(2) \times U(1)$ is valid only for massless gauge bosons and massless fermions. Fermions must be massless because the mass terms in the Lagrangian $m\bar{\psi}\psi$ couple left-handed isodoublets and right-handed isosinglets ($\bar{\psi}\psi = \bar{\psi}_L\psi_R + \bar{\psi}_R\psi_L$) and thus violate the conservation of both isospin and hypercharge. Not only the local, but also the global symmetry $SU(2) \times U(1)$, are thus broken in nature.

The so-called standard electroweak theory is based on the assumption that this breaking of the $SU(2) \times U(1)$ symmetry takes place spontaneously. It is the mechanism of this symmetry breaking that is the ugly duckling of the theory.

When discussing the approximate global chiral symmetry of QCD in Chapter 3, we said that spontaneous breaking of global symmetry produces massless Goldstone bosons. The spontaneous breaking of local symmetry produces an effect that, in some loose sense, is the opposite. Massless gauge fields become massive, "eating up" the unborn Goldstone bosons. A massless vector field with two spin states and a massless scalar field give birth to a massive vector particle with three spin projections. The number of degrees of freedom is conserved. This effect was discovered in field theory in 1964 and is called the Higgs mechanism.

A concrete realization of the Higgs mechanism in the standard theory of the electroweak interaction is based on the use of an isotopic doublet of scalar particles, φ^+, $\varphi^°$, and, of course, of the corresponding antiparticles, φ^-, $\bar{\varphi}^°$. These scalar fields have both isospin and hypercharge and interact in a gauge-invariant way with the four gauge fields, W^+, W^-, $W^°$, and $B^°$. The corresponding term in the Lagrangian is

$$|D_\mu \varphi|^2 = (D_\mu\varphi)_i^* (D_\mu\varphi)_i,$$

where the covariant derivative

$$D_\mu = \frac{\partial}{\partial\mu} - ig_2 \frac{\vec{\tau}}{2} \vec{W}_\mu - ig_1 \frac{Y}{2} B_\mu$$

acts on the isotopic spinor $\varphi = \binom{\varphi^+}{\varphi^°}$ and a summation is carried over the isotopic suffix $i = 1,2$.

The scalar field φ also interacts with fermions, again conserving the isotopic spin and hypercharge. This interaction transforms isosinglet right-handed fermions into isodoublet left-handed ones. Such fermion-scalar interactions (usually called Yukawa couplings) exist for all six lepton and quark pairs; two for each pair if neutrinos are regarded on the same footing as other particles, without assuming them to be massless. So far we know of no theoretical principle for choosing Yukawa coupling constants. This arbitrariness looks very unattractive.

Both the gauge and the Yukawa interactions of scalar particles in $SU(2)\times U(1)$ are locally invariant and do not directly produce the spontaneous breaking of the $SU(2)\times U(1)$ symmetry. This breaking stems from the nonlinear interaction between the fields φ that we shall write in the form of a potential (see Figure 31),

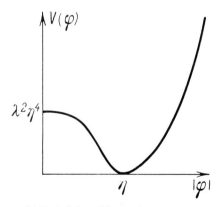

FIGURE 31. The potential $V(\varphi)$ of the self-interacting scalar field with a minimum at $|\varphi| = \eta$.

$$V(\varphi) = \lambda^2(|\varphi|^2 - \eta^2)^2,$$

where λ is a dimensionless parameter whose value is not yet known. The quantity

$$|\varphi|^2 = \bar\varphi_i \varphi_i = \bar\varphi^+ \varphi^+ + \bar\varphi^\circ \varphi^\circ$$

is an isoscalar. The parameter η has the dimension of mass. In order to obtain the correct value of the Fermi constant G_F, we have to choose (see below)

$$\eta = 2^{-3/4} \times G_F^{-1/2} = 174 \text{ GeV}.$$

Since the constants of gauge interactions, g_1 and g_2, and the constants of Yukawa interactions are dimensionless (in the $\hbar = c = 1$ units), the parameter η is the only dimensional parameter of the theory. The masses of all the particles are expressed through η.

The main difference between the outlined theory and those discussed previously lies in the minus sign in the expression for $V(\varphi)$. This minus sign is critical. Were there a plus sign instead of a minus in this expression, the theory would be stable, no spontaneous breaking of symmetry would take place, and the vector bosons and fermions would remain massless. As a result, the theory would not be able to describe the real world in which these particles are massive.

We already know that, in the case of a spontaneous breaking of symmetry, the Lagrangian possesses a symmetry while the physical states do not. In particular, the vacuum, i.e., the physical state with minimum energy, does not have the symmetry of the Lagrangian. In our case, the vacuum is not symmetric because of the above minus sign. Indeed, the expression for $V(\varphi)$ immediately shows that the potential energy is zero when $|\varphi| = \eta$. This means that a constant scalar field, the so-called vacuum condensate of the scalar field, exists in the vacuum. But since the scalar field φ possesses both isospin and hypercharge and is noninvariant under this group SU(2) × U(1), the vacuum is also noninvariant under this group. The symmetry is found to be spontaneously broken.

By using the isotopic symmetry of the original Lagrangian, we can choose the field φ in such a manner that the vacuum expectation value will be nonzero only for the lower, electrically neutral, component of the isotopic spinor,

$$<\varphi> = \begin{pmatrix} 0 \\ \eta \end{pmatrix}.$$

Consequently, the existence of the condensate does not violate the conservation of electric charge.

Let us now see how the condensate imparts masses to intermediate bosons. To understand this, we shall look more closely at the expression for $|D_\mu \varphi|^2$ given above. By substituting in it the vacuum expectation value $<\varphi> = \binom{0}{\eta}$ for φ, we immediately obtain the term giving mass to W^+ bosons

$$\frac{g_2^2}{2} \eta^2 \overline{W} W ,$$

whence $m_W = g_2 \eta / \sqrt{2}$. In order to obtain the mass term of the Z boson, we must use the central formula (see p. 66). The result is

$$\tfrac{1}{4}(g_1^2 + g_2^2) \eta^2 ZZ,$$

whence $m_Z = m_W / \cos \theta_W$.

Since the condensate $<\varphi>$ is electrically neutral, the operator of charge Q applied to it gives zero and the photon remains massless. The intermediate bosons have thus become massive because the isospin is not conserved, that is, the condensate has isospin; and the photon remains massless because electric charge is conserved owing to the neutrality of the condensate.

The relation between m_W and η established above makes it possible to determine η by comparing the two relations

$$m_W = g_2 \eta / \sqrt{2} \quad \text{and} \quad \frac{G_F}{\sqrt{2}} = \frac{g_2^2}{8 m_W^2}$$

This immediately gives a relation between η and G_F that does not contain g_2,

$$\eta = (2^{3/2} G_F)^{-1/2} = 174 \text{ GeV}.$$

This formula has already been mentioned in the discussion of the nonlinear potential $V(\varphi)$ of scalar fields.

Let us now see how fermions become massive. The sources of their masses are Yukawa couplings of the type

$$f(\bar{\psi}_L \psi_R \varphi + \bar{\psi}_R \psi_L \bar{\varphi}).$$

When the scalar field φ gets the vacuum expectation η, a fermion is given the mass $m = f\eta$.

The values of the various Yukawa constants f have to be very small (from 10^{-6} to 10^{-1}) in order to reconstruct the mass spectra of leptons and quarks. No symmetry can be suggested at present to govern their values. Attempts have been made to construct the hierarchy of these constants in powers of α in perturbation theory, but so far they have not gone far. . . .

HIGGS BOSONS

Among the four scalar fields, φ^+, φ°, φ^-, and $\tilde{\varphi}^\circ$, three are "eaten up" by vector bosons, as described above, as a result of spontaneous symmetry breaking. Only one neutral scalar field survives to represent "live" scalar bosons: quanta of scalar waves against the background of constant condensate η,

$$\varphi = \begin{pmatrix} 0 \\ \eta + \chi \end{pmatrix}.$$

By substituting this expression for φ into the expression for the potential $V(\varphi)$, it is not difficult to find the mass of these so-called Higgs bosons: $m_H = 2\lambda\eta$. We cannot predict the masses of Higgs bosons because the constant λ is unknown. A theoretical analysis shows that they cannot be very light: The minimum value of m_H is in the vicinity of 7 GeV. Neither can they be superheavy in the framework of the approach described, but a mass of the order of 1 TeV cannot be excluded.

It must be clear by now that the heavier a particle, the stronger its interaction with a Higgs boson; indeed, Higgs bosons give masses to other particles, the masses being the larger, the stronger the interaction with other particles is. For instance, in tt̄ quarkonium with a mass of around 50 GeV, the ratio of decays through the channels Hγ and $\mu^+\mu^-$ will be approximately 1:10 and the Hγ decay will not be difficult to find experimentally, provided that the H bosons are relatively light. Decays of such light H bosons will mostly produce hardrons containing heavy quarks, that is, bb̄ and cc̄, and $\tau\bar\tau$ lepton pairs.

Reactions involving W and Z bosons are especially promising in the search for H bosons. For instance, a light H boson can be found among LEP events in reactions of the type

$$e^+e^- \rightarrow Z^* \rightarrow HZ \qquad \text{or } e^+e^- \rightarrow Z \rightarrow HZ^*$$
$$\hookrightarrow e^+e^- \qquad\qquad\qquad \hookrightarrow e^+e^-,$$

where Z^* is a virtual Z. H bosons must accompany the creation of W and Z bosons on pp̄ colliders (with a probability of about 10^{-3}).

The heavier an H boson, the more difficult it will be to create it in a collider, because the required energy must be correspondingly greater. However, once a heavy H boson has been created, it will be easier to detect than a light one. Indeed, the decay products of heavy H bosons must have high transverse momenta. If the mass of the H boson exceeds 180–200 GeV, it will readily decay into $Z°Z°$ and W^+W^- pairs. This would be a beautiful phenomenon.

MODELS, MODELS . . .

The breaking of the $SU(2) \times U(1)$ symmetry that is realized by the doublet of scalar fields and gives life to one scalar neutral Higgs boson, is only one of a multitude of theoretical versions of the way the electroweak theory is broken. Papers have been published on models with several Higgs bosons, including charged bosons.

In some models, attempts are made to obtain the mirror asymmetry (which is inserted "by hand" into the foundations of the $SU(2) \times U(1)$ symmetry) by a spontaneous breaking of a more general mirror-symmetric theory of the type $SU(2)_L \times SU(2)_R$ that contains both left- and right-handed fermion doublets. This symmetry also contains two types of gauge fields: "left-handed" and "right-handed" intermediate bosons. A spontaneous breaking of symmetry gives greater masses to "right-handed" bosons than to "left-handed" bosons. Right-handed currents are therefore coupled less strongly than left-handed currents. It would be extremely interesting to look experimentally for the right-handed currents predicted by such models. We shall remark at this juncture that the accuracy to which the absence of right-handed charged currents has been verified experimentally is not better than 1 percent (such is the accuracy achieved in measuring, for example, the longitudinal polarization of electrons in the β decay).

Speaking of multiple-Higgs models, we have to mention certain attempts to "hide" the source of CP noninvariance in the Higgs sector of the theory, namely, in the terms of the Lagrangian describing the interaction between different Higgs bosons. Such theoretical models predict a relatively high dipole moment for the neutron, $d_n \sim e \cdot 10^{-25}$ cm, close to the experimentally obtained upper bound. The same models also predict a relatively large difference between dimensionless amplitudes describing the CP-odd decays $K°_L \to \pi^+\pi^-$ and $K°_L \to \pi°\pi°$,

$$|\eta_{+-} - \eta_{00}| : |\eta_{+-}| \sim 6 \text{ percent}$$

(the current experimental estimate of this ratio is 3 ± 4 percent).

Many theorists believe that the potential

$$V(\varphi) = \lambda^2(|\varphi|^2-\eta^2)^2$$

is too artificial. They prefer to begin with a stable potential with a plus sign in front of the term η^2 or, better still, with $\eta=0$. Interestingly, instability and spontaneous symmetry breaking can occur in this case as well. But they appear only after we take the induced interactions between scalar bosons into account, that is, the interactions due to radiative corrections: loops formed by virtual gauge fields. The effective potential appearing in this case (the Coleman-Weinberg potential $\sim|\varphi|^4\ln(|\varphi|^2/m^2)$) has a minimum at $|\varphi|^2 \neq 0$ and, hence, generates a scalar condensate.

The presence of fundamental scalar particles is common for all the above-discussed models of the breaking of the electroweak symmetry. Attempts to purge these particles from a theory have demonstrated that, once we rid the model of funadmental scalar bosons, composite scalar bosons invariably appear. Furthermore, the constituents of these bosons must be confined over such small distances ($\leq 10^{-17}$cm) that the composite nature of the bosons cannot be revealed at energies available at present, and they should appear as pointlike particles.

The models with composite scalars are called technicolor models. They postulate the existence of a large number of new particles: so-called techniquarks and technigluons with a confinement radius of the order of 10^{-17}cm. They give masses to the W and Z bosons, eating up Goldstone technipions that appear as a result of a spontaneous breaking of chiral symmetry in quantum technichromodynamics. Unfortunately, technicolor offers no natural mechanism for giving masses to fermions, and this part of the model looks rather unattractive.

As far as technicolor is concerned, I would like to add that, to me, fundamental scalars do not seem less attractive than fundamental vector or scalar fields. I do not share the antipathy of many theorists to fundamental scalars. If we choose to make Higgs bosons composite, then it is natural to have composite quarks, leptons, intermediate bosons, and even massless gauge fields, that is, gluons and photons. But this is quite a different topic.

SCALARS: PROBLEM NO. 1

Whatever the attitude to specific models discussed in the preceding section, one thing is obvious: We cannot do without scalar particles. They are connected to the most fundamental of the yet-unsolved problems of elementary particle physics: the problem of masses and the related problems of quark mixing in weak currents, CP violation, and, possibly, P violation.

While vector fields incarnate dynamics, scalar fields incarnate inertia. While vector fields are a striking manifestation of (local) symmetry, scalar fields play a no less important part: symmetry breaking. After intermediate bosons have been discovered and the fundamental gauge ideas in high energy physics have been thereby confirmed, no problem will be more important than the discovery of scalar bosons and the study of their properties.

Theoretical physicists penetrated into scalarland by trying to construct a working, viable model of the electroweak interaction. But this land is interesting in itself as well. As Bjorken suggested, if you strike out all gauge charges in the fundamental Lagrangian, you will have on your hands a large number of terms about whose nature and symmetry properties we can only guess at present. Numerous theoretical models discussed in the literature are forerunners of the birth of a new physical continent: the physics of fundamental scalars.

ON THE DEVELOPMENT OF THE THEORY

What can be said about the part played by theorists in the experimental discoveries of the physics of elementary particles?

We can name a number of discoveries that were completely unexpected by the theorists. For instance, radioactive decay, the muon, strange particles, and CP violation were not predicted. Moreover, a completely satisfactory answer to the question "Who ordered the muon?" has not yet been found. Unexpected discoveries played an extremely important role in the history of physics. It can be safely predicted that the inflow of such surprises will not peter out in the future. Indeed, our knowledge still covers a very small region compared to the region of the unknown. And from each new big accelerator physicists expect unexpected discoveries.

There have also been discoveries in which theorists' predictions were of paramount importance. Among the examples we can point to are Pauli's hypothesis of the existence of neutrino, which waited more than a quarter of a century for a direct experimental test, or the brilliant list of experiments searching for the parity violation in weak interactions suggested by Lee and Yang. In both these cases, the prediction was preceded by a profound theoretical phenomenological analysis of the results of a series of experiments that led to a seemingly unsolvable paradox. In the case of the neutrino, this was the leakage of energy in the β decay; in the case of parity, it was the θ–τ paradox (at that time θ denoted (K) mesons decaying into two pions and τ denoted (K) mesons decaying into three pions). But, having resolved these paradoxes, theorists only made inescapable logical conclusions from experimental data. One could even say that they simply had no choice.

Finally, we know of discoveries of the third type, in which the role of the theorist is exceptional. These discoveries mostly stem from the internal development of theoretical physics per se. Pertinent examples are the deflection of light rays by the gravitational field of the sun, the positron, and neutral currents. The origins of these predictions were far away (although to a different extent) from the hot problems of experimental physics. The bending of light rays by the Sun was calculated by Einstein in accord with the postulated invariance of the equations of general relativity under local coordinate transformations of the most general type. The prediction of the positron was made as a result of the "crossing" of quantum mechanics with special relativity by Dirac. The prediction of neutral currents was a corollary of the electroweak theory. Among the corollaries of this theory are the W and Z bosons (I hope that they will have been definitely discovered by the time this book is printed)† and the Higgs bosons.

Of course, experimental discoveries played a tremendous role in the concrete realization of the ideas of the electroweak theory. This is especially true in the establishment of the universality of the weak interaction and its mirror asymmetry. But the main driving force in the progress of this theory was, nevertheless, purely theoretical.

The idea of intermediate vector bosons was formulated by Yukawa as early as 1935, and theorists returned to it more than once. This idea was attractive because the interaction of bosons with fermions is characterized by a dimensionless constant and is thus described by a renormalizable theory that, in principle, enables one to carry out calculations in the higher orders of perturbation theory and to express the results of these calculations in terms of several experimentally measured parameters, such as the charges and masses of particles. As far as computability is concerned, this theory sharply differs from the four-fermion theory, in which each new order of perturbation theory introduces fresh, more frightful, and more numerous divergencies. In fact, the four-fermion Lagrangian gives meaningful answers only in the first order of perturbation theory in G_F, and even then it is not clear why.

---

†See Author's note to p. 113.

THE ELECTROWEAK THEORY

Actually, it was found, even before World War II, that massive vector bosons also lead to divergencies sufficient to destroy renormalizability, although not as strong as those given by the four-fermion interaction. However, purely theoretical discoveries—first of the local isotopic symmetry† and Yang–Mills field, then of the Higgs mechanism—made it possible to get rid of this defect of vector bosons and to construct a renormalizable theory.

Looking back, we recognize that theorists, advancing step by step, injecting new ideas related to symmetries, were constructing the edifice of an internally consistent theory, a theory that would enable one to carry out calculations and expect meaningful results with a minimum number of external parameters.

At the present moment, scalar bosons are not directly required to explain any available experimental facts. The theory needs these bosons only to remove certain remaining divergencies in higher orders of perturbation theory. If we are sure that scalar bosons exist, then this feeling is ultimately based on the ideas of computability, theoretical beauty (symmetry), and self-consistency.

When thinking about the history of elementary particle physics, one is tempted to imagine that symmetry itself, breaking through the asphalt crust of human lack of understanding, guides a theorist's pen and thus persuades an experimentalist to discover its existence.

In the next chapter, we shall see quite a few examples of astonishing "self-conception" and subsequent development of symmetry ideas that led to the prediction of a number of striking phenomena, such as proton decay, magnetic monopoles, and numerous supersymmetric particles. It would be remarkable if even some of these theoretical dreams came true.

---

†The first attempt to employ local isotopic symmetry for a description of the weak (and strong) interaction was made by O.Klein as early as 1938. Klein treated the doublet of nucleons (n,p) and the doublet of leptons (ν,e) on an equivalent basis. One of the versions of his theory contained four gauge bosons: the photon and (in today's notation) W^+, W^-, and Z°, with all gauge interactions characterized by a single constant, namely, electric charge e.

In constructing this theory, Klein introduced a hypothesis of the existence of a fifth dimension. The model of a five-dimensional world with a cyclic fifth coordinate (the so-called Kaluza-Klein model) was the object of a large number of studies in the period between the two World Wars. Klein attempted to construct, in the framework of this model, a unified theory of the electromagnetic, weak, and gravitational interactions. Unfortunately, Klein failed to make the simplest step: he did not evaluate the expected masses of the W and Z bosons. Klein's paper had been forgotten; thus, modern gauge theories originate with the work of Yang and Mills.

It should be noted in this connection that we have witnessed in recent years a revival of acute interest in theories with extra spatial dimensions.

Chapter 6

PROSPECTS FOR UNIFICATION

Rendezvous of the running coupling constants
Fermions in SU(5)
Gauge bosons in SU(5)
Proton decay
Magnetic monopoles
Models, models, models . . .
Supersymmetry
Unification models and the Big Bang
On extrapolations and predictions

RENDEZVOUS OF THE RUNNING COUPLING CONSTANTS

If one forgets for a moment the "haunting problems" concerning the particles' masses and scalar bosons, the existing picture of fundamental forces looks very beautiful: strong, weak, and electromagnetic interactions are caused by the local symmetry group $SU(3) \times SU(2) \times U(1)$ with its three coupling constants, or "charges", g_3, g_2, and g_1, and twelve gauge fields: eight gluons, three intermediate bosons, and the photon. At sufficiently short distances, all these forces resemble one another and give a Coulomb-type potential: $\sim g^2/r$. In the case of the strong interaction, the term "short distances" means lengths much smaller than the size of hadrons, that is, much smaller than 10^{-13}cm; where asymptotic freedom reigns. In the electroweak interaction, Coulomb's law sets in at distances much shorter than the Compton wavelength of the W and Z bosons, that is, 10^{-16}cm. At such short distances, the fact that bosons are massive is unimportant.

The three charges are not very different at distances of about 10^{-17}cm:

$$\alpha_s \equiv \frac{g_3^2}{4\pi} \approx \frac{1}{10} \ , \ \alpha_w \equiv \frac{g_2^2}{4\pi} \approx \frac{1}{27} \ , \ \alpha_{em} \equiv \frac{e^2}{4\pi} \approx \frac{1}{129}$$

Note that α_{em} increases in comparison with the standard macroscopic value (1/137) because vacuum screening diminishes with diminishing distance. We mentioned at the beginning of this book that, owing to the effect of vacuum polarization, α_s and α_w decrease as momentum transfer increases, while α_{em} increases.

Figure 32 shows the logarithmic plot of the reciprocal coupling constants $1/\alpha_s$, $1/\alpha_w$, and $\frac{3}{8} (1/\alpha_{em})$ as functions of momentum transfer q measured in GeV (the explanation for the coefficient 3/8 will be given several pages later).

According to the theory, $1/\alpha_i$ is an approximately linear function of log q. The $1/\alpha_s$ curve has the largest slope; it is caused by the polarization of the gluon vacuum. This slope is greater than that of the $1/\alpha_w$ curve because gluons, being more numerous than intermediate bosons, produce a larger antiscreening effect (the greater the number of gauge fields the stronger the tendency to asymptotic freedom). In contrast to this, screening predominates for α_{em}; thus, $1/\alpha_{em}$ diminishes as q increases.

Figure 32 shows that the reciprocal constants $1/\alpha_i$ converge to the rendezvous value $1/\alpha_{GU} \simeq 40$, at which $q_{GU} \simeq 10^{14}$–10^{15} GeV. The subscript GU stands for the grand unification of the three fundamental interactions.

It seems natural to assume that the electromagnetic, weak, and strong interactions are indistinguishable not only at $q = q_{GU}$, but also at $q > q_{GU}$ and are described by a unifying simple local symmetry group with a common gauge coupling constant α_{GU} (see Figure 32).

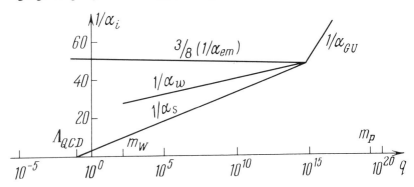

FIGURE 32. The running coupling constants of the electromagnetic, weak, and strong interactions.

A simple minimum-rank group that contains the product SU(3)×SU(2) ×U(1) as a subgroup is the group SU(5). It was suggested as the symmetry of grand unification by Georgi and Glashow in 1974.

FERMIONS IN SU(5)

We begin the description of the SU(5) symmetry with fermions, considering first, for the sake of simplification, only one generation. A spinor transforming under the fundamental representation of the group SU(5) is the five-component spinor containing five massless left-handed particles: three anti-d-quarks differing in colors ($\tilde{d}_{1L}, \tilde{d}_{2L}, \tilde{d}_{3L}$), the electron e_L, and the neutrino ν_L (remember that a left-handed particle has negative helicity: its spin points against its momentum, $\psi_L = \frac{1}{2}(1+\gamma_5)\psi$). In further discussion, we deal primarily with left-handed particles, and thus sometimes drop the subscript L.

The fields e_L and ν_L form a doublet under the group SU(2) and are color singlets; $\tilde{d}_{1L}, \tilde{d}_{2L}, \tilde{d}_{3L}$ form an (anti) triplet under the group SU(3) and are singlets under weak SU(2).

Putting leptons and quarks into one SU(5) multiplet explains why the charge of the d quark is 1/3 of the electron charge. By postulating that all interactions be described by the group SU(5), we included the electric charge Q (or rather, a quantity Q' proportional to Q) into the set of generators of the group. This, in turn, imposes on the sum of the charges of the particles belonging to a given SU(5) multiplet the requirement that it be equal to zero. (Otherwise, the charge of each particle would have an SU(5) invariant component and, therefore, the charge would not be a generator of the group SU(5). Compare this to the formula $Q = T_3 + Y/2$ for the group SU(2) and to the nonzero hypercharge of the isospinor $\nu_L e_L$.) This means that

$$3Q_{\tilde{d}} + Q_e + Q_\nu = 0.$$

Since $Q_\nu = 0$ and $Q_{\tilde{d}} = -Q_d$, we obtain $Q_d = \frac{1}{3}Q_e$. Let us now turn to other fermions of the first generation. The left-handed antineutrino not participating in the electroweak interaction is an SU(5) singlet. The remaining ten left-handed fermions, $3d_L$, $3u_L$, $3\tilde{u}_L$, and e^+_L, form an SU(5) decouplet.

The particles of the other two generations are distributed among the SU(5) multiplets in the same manner (note, in this connection, that the upper quarks in the SU(5) multiplets must obviously be rotated).

D

GAUGE BOSONS IN SU(5)

The group SU(2) has three gauge bosons, SU(3) has eight, and SU(5) has 24. Twelve bosons among these 24 are our good acquaintances: 8 gluons, 3 intermediate bosons, and the photon; the other twelve bosons are new.

By comparing interactions between the "old" bosons, we can give here the promised explanation of the factor 3/8, without which α_{em} could not be in time for the meeting with α_s and α_w (see Figure 32).

Remember that the interaction between gluons and fermions is proportional to $g_3 \lambda/2$ and the interaction of W bosons and fermions is proportional to $g_2 \tau/2$, where λ are the eight Gell-Mann matrices and τ the three Pauli matrices. The constants thus defined, g_3 and g_2, are identically normalized because the matrices τ and λ have the same normalization (in fact, the matrices, λ_1, λ_2, and λ_3, coincide with τ_1, τ_2, and τ_3). As for photons, their interaction is proportional to eQ.

The operator of the charge Q is not a generator of the group SU(5). The generator is a quantity $Q' = cQ$, where the coefficient c is easily found from the requirement of identical normalization of Q' and, say, the isospin operator $\tau/2$. Normalization is carried out most simply by using the five-component spinor, $\tilde{d}_{1L}, \tilde{d}_{2L}, \tilde{d}_{3L}, e^-_L$, and ν_L. The sum of the squares of the isospin projections of the particles, $\sum T_3^2$, is $\frac{1}{4} + \frac{1}{4} = \frac{1}{2}$ (the isospin of the $\tilde{d}_L$ quarks is zero, $T_3 = +1/2$ for the neutrino, and $T_3 = -1/2$ for the electron). Of course, the sum of the squares of the normalized charges Q'^2 must also equal 1/2. But

$$\sum Q'^2 = c^2 \sum Q^2 = c^2(3(\tfrac{1}{3})^2 + 1 + 0) = \tfrac{4}{3}c^2;$$

hence, $c^2 = 3/8$. If the interactions with the photon are written in the form $e'Q' = eQ$, we find that $e' = \sqrt{8/3}\, e$. The running constants α_s and α_w must be compared not with $\alpha_{em} = e^2/4\pi$, but with $\alpha'_{em} = (e')^2/4\pi = \tfrac{8}{3}\alpha_{em}$; it is this comparison that Figure 32 demonstrates.

Look at the five-component spinor. Quarks of different colors transform into one another, emitting and absorbing gluons. The electron and the neutrino transform into each other by emitting and absorbing W bosons. And what are the transitions connected with the emission and absorption of the twelve new gauge fields? The answer is obvious: six of them mediate transitions between the $\tilde{d}$ quarks and the electron. Their charges are $\pm 4/3$. These are the so-called X bosons. The other six, the so-called Y bosons, mediate transitions between the $\tilde{d}$ quarks and the neutrino; their charges are $\pm 1/3$.

The masses of X and Y bosons must be of the order of the grand unification energy: $m_X \approx m_Y \approx 10^{14} - 10^{15}$ GeV. If $q \gg m_X, m_Y$, then the symmetry is SU(5). If $q < m_X, m_Y$, then the SU(5) symmetry is broken. It is assumed that SU(5) is broken spontaneously by very heavy Higgs fields whose masses are close to m_X and m_Y.

PROTON DECAY

If the X and Y bosons interacted only with the five fermions $\tilde{d}, e, \nu$, the baryon charge would be conserved, in spite of the fact that these bosons transform quarks into leptons. The situation would be similar to that realized by the W bosons. These last produce the electron-neutrino transitions but certainly conserve electric charge. A virtual W boson removes the charge from fermions when it is emitted and returns it to them when absorbed.

Nonconservation of baryon charge in the case of the X and Y bosons occurs because these bosons simultaneously interact with currents of two different types. The reason is that the number of elementary currents is much larger than the number of bosons. Indeed, the total number of gauge bosons is 24, while the number of current transitions among the 15 fermions of one generation (we ignore the sterile $\tilde{\nu}_L$) is 125. Hence, the second job of X bosons is to transform u quarks into ū quarks, and that of Y bosons is to transform u quarks into $\tilde{d}$ quarks. The same boson can thus transform both into an antilepton-antiquark pair and into a pair of quarks,

$$uu \leftarrow X \rightarrow e^+ \tilde{d},$$
$$ud \leftarrow Y \rightarrow \tilde{\nu}_e \tilde{d}.$$

By reversing the left-hand arrows and transferring $\tilde{d}$ from the right- to the left-hand side (in this operation, the creation of $\tilde{d}$ is, as always, replaced with the annihilation of d), we obtain the transitions

$$uud \rightarrow e^+,$$
$$udd \rightarrow \tilde{\nu}.$$

Recall now that the combination uud is the proton and udd is the neutron. We have thus obtained the processes of decay of nucleons. Obviously, emission of a single lepton is forbidden by the laws of energy and momentum conservation, and so these decays are of the type

$$p \rightarrow e^+ \pi^°, \ n \rightarrow \tilde{\nu}_e \pi^°, \ n \rightarrow e^+ \pi^-, \ p \rightarrow e^+ \pi^+ \pi^-,$$

and so on.

The matrix element of the proton decay must be of the order of $\alpha_{GU} m_X^{-2}$, and the decay rate must be of the order of $\alpha_{GU}^2 m_X^{-4} m_p^5$. Here $\alpha_{GU} \approx 1/40$ is the grand unification constant, $m_X = 10^{14}$–10^{15} GeV is the X boson mass, and m_p is the proton mass, which enters the resultant expression because the energy release in the decay is approximately equal to m_p. The power of m_p must be five, because the dimension of decay rate per unit time equals the dimension of mass (recall the $G_F^2 \Delta^5$ law for weak decays).

Substituting $m_X/m_p = 10^{14}$ into the above estimate for the decay rate and converting to conventional units, we obtain $\tau_p = 3 \times 10^{27}$ years for the proton lifetime. With the ratio $m_X:m_p$ increased by one order of magnitude, we find that $\tau_p = 3 \times 10^{31}$ years. Exact calculations indicate that the proton lifetime falls into this interval. Assuming that the symmetry realized in nature is the so-called minimal SU(5) symmetry (i.e., a symmetry with a minimal number of Higgs fields), we find that the main uncertainty of these calculations lies in the choice of Λ_{QCD}, since it determines the point on the abscissa in Figure 32 through which the $1/\alpha_s$ curve passes.

Figure 32 shows that m_X is proportional to Λ_{QCD} and, hence, τ_p is proportional to Λ_{QCD}^4. A time of the order of 10^{28} years corresponds to $\Lambda_{QCD} \approx 100$ MeV. When Λ_{QCD} is "shifted", not only m_X, but also α_{GU} and α_w (m_w^2) are forced to shift; hence, the value of $\sin^2 \theta_w$ predicted by the SU(5) theory shifts as well. Theoretical predictions of the value of $\sin^2 \theta_w$ group around the value 0.22, in good agreement with experimental data.

Figure 32, with its logarithmic scale along the abscissa axis, somehow conceals the scale of extrapolation by 14 orders of magnitude in energy covered by the grand unification model. It is indeed remarkable that, by studying the interaction between neutrinos and hadrons and the e^+e^- annihilation into hadrons, one can, through accelerator experiments, draw conclusions about physics at distances of the order of 10^{-28} cm!

Although the interval of 10^{30} years is approximately 20 orders of magnitude greater than the lifetime of the universe, two reasons make the measurement of such an unimaginably long life of the proton quite feasible. First, by virtue of the laws of quantum mechanics, the decay of even one nucleon can be observed over any, no matter how short, time t, but with a low probability, t/τ_p. Second, there are plenty of nucleons around us: 6×10^{23} nucleons per gram of any substance. Hence, if we take, say, 16 tonnes of water (i.e., 10^{31} nucleons) and observe this mass for a year, we expect to detect ten decays of protons and neutrons if the nucleon lifetime is 10^{30} years.

The main problem with such an "observational experiment" is the background due to cosmic rays. Indeed, one high-energy particle arrives each second per cm² of the Earth's surface. To reduce this background, the sample must be placed deep underground. This drastically diminishes the flux of the charged particles, so that the background is mostly caused by the neutrino flux. This background is very low because neutrinos interact with matter very weakly; however, the effect we are after is still smaller. In order to detect the proton decay reliably, we have to check the balance imposed by the conservation of energy and momentum.

The search for the proton decay is being conducted or prepared at present in at least twenty underground laboratories. At the moment of writing it is reliably established that the proton lifetime is greater than 10^{31} years.

The scientific community was very much stirred when some experimenters reported that events classifiable as "candidates" for proton decay were observed. Several such cases were reported by scientists working in the deepest mine in the world, in India. One candidate was reported by a group of physicists working in the tunnel under Mont Blanc. If these cases were indeed proton decay events, its lifetime would be around 5×10^{30} years. With this lifetime, the probelm of proton instability could be finally solved in the very near future, when multikiloton underground detectors are put into operation.†

If the proton lifetime proves to be in the range 10^{30}–10^{32} years, we should consider ourselves extremely lucky. In fact, not one of the constructed detector setups is capable of recordng the decay if $\tau_p > 10^{33}$ years. Many scientists are of the opinion that, if $\tau_p > 10^{35}$ years, the decay would be so infrequent as to be practically undetectable. In any case this would call for fantastically large detectors.

The grand unification mass $m_X \cong 10^{14}$–10^{15} GeV is much closer to the Plank mass $m_P \cong 10^{19}$ GeV than to the masses and energies with which physicists deal in their accelerator experiments. If grand unification is indeed the reality, there undoubtedly must be a relationship between m_X and m_P. The discovery of the proton decay would be the discovery of the century. As a musician's fork, it would tune the whole physics of elementary particles to the Plank frequency and predetermine its course for many years to come.

---

†*Author's note (autumn, 1983).* The Irvine-Michigan-Brookhaven collaboration, working with an 8000 tonne Cherenkov water detector placed at a depth of 1570 m of water equivalent, has established a lower limit $\tau_p/B\,(p \to e^+ \pi^\circ) > 6.5 \times 10^{31}$ years. Here τ_p is the proton lifetime and $B\,(p \to e^+ \pi^\circ)$ is the corresponding branching ratio for the decay channel $p \to e^+ \pi^\circ$. This result eliminates the SU(5) model with the minimum number of particles but does not exclude more complex versions of grand unification models.

MAGNETIC MONOPOLES

A magnetic monopole, that is, a magnetic charge, is known to be the source of a spherically symmetric magnetic field whose strength diminishes as $1/r^2$. No magnetic monopoles have been experimentally discovered thus far.

Grand unification models based on compact semisimple and simple gauge groups (among the latter is SU(5)) have magnetic monopoles as solutions. These monopoles are very heavy: their masses are of the order of $m_\sqrt{}/\alpha_{GU}$, that is, of the order of 10^{16} GeV. The nature of these monopoles is essentially different from the nature of other elementary particles. A monopole is a nontrivial topological construction extended in space; it is built of non-Abelian scalar and gauge fields.

Monopole solutions in non-Abelian gauge theories were first constructed by 't Hooft and Polyakov in 1974. They have found the following solution for the group SU (2) with a triplet of scalar Higgs fields φ^a ($a=1, 2, 3$) and a triplet of gauge vector fields A_μ^a ($\mu = 0, m$ where $m = 1, 2, 3$):

$$\varphi^a = c\delta_{an}\frac{x_n}{r}H(r),$$

$$A_m^a = \epsilon_{man}\frac{x_n}{2er^2}F(r) \; , \; A_0^a = 0.$$

Here e is the gauge charge; $a=1, 2, 3$ are the indices of coordinates in isotopic space; $m,n =1, 2, 3$ are the indices of the coordinates $\mathbf{x}$ in ordinary three-dimensional space; $r^2 = \mathbf{x}^2$; c is a coefficient with the dimension of mass; $H(r)$ and $F(r)$ are dimensionless functions that equal zero when $r=0$ and rapidly tend to 1 when $r \gg 1/c$, that is, outside the core of the monopole; and $\delta_{am} = 1$ if $a = m$ and $\delta_{am} = 0$ if $a \neq m$. Furthermore, $\epsilon_{amn} = \epsilon_{mna} = \epsilon_{nam} = +1$ and $\epsilon_{anm} = \epsilon_{nma} = \epsilon_{man} = -1$ if a, m, and n are distinct and $\epsilon_{amn} = 0$ if any two of the indices are identical. By convention, repeated indices denote summation.

This solution has a spectacular property. The point is that the orientation of the isotopic spin of the Higgs and gauge fields in isotopic space is different for different points of ordinary space. Note that the isotopic vector of the Higgs field in isotopic space points in the same direction as the radius vector in ordinary space, while the isotopic vector of the gauge field is orthogonal to the radius vector.

Polyakov christened this solution a "hedgehog." By employing the local isotopic invariance of the theory, we can try to "comb" the hedgehog, directing the isotopic vector of the classical Higgs field (condensate) identically throughout the space, say, along the third axis. Under the standard definition of electric charge, the Higgs condensate will then be explicitly electrically neutral. It will give masses to charged vector fields, but will leave the neutral vector field, that is, the photon, massless. In contrast to what we have in the electroweak theory, such a Higgs condensate would vanish in a small neighborhood of the origin of coordinates ($r \lesssim 1/c$). Consequently, all three vector fields are massless at such short distances. Correspondingly, the gauge SU(2) symmetry does not break within the monopole.

The reader will readily notice (see Figure 33) that the hedgehog cannot be combed by a continuous transformation. Figure 33 shows the presence

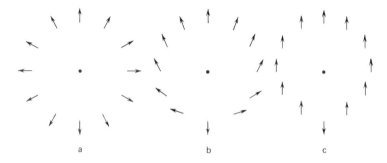

FIGURE 33. The field of a non-Abelian monopole at a given distance from its center; the arrows point in the direction of the field in the isotopic space. (a) Uncombed monopole, (b) monopole being combed, (c) monopole combed everywhere except at a single point ("South Pole").

of a singularity along the semiaxis from the center toward the south pole. It can be shown that this singularity corresponds to the Dirac string: an infinitely long thin unobservable solenoid supplying a magnetic field flux into the monopole, from which it is "sprayed" through the whole space, with spherical symmetry. The magnetic field is then shown to be $\mathbf{H} = \mu \mathbf{x}/r^3$, where $\mu = 1/e$ is the magnetic charge. If we take into account the fact that, in such an SU(2) theory, the charge will be minimal for particles with charges $+e/2$ and $-e/2$ that form a hyperneutral isotopic doublet, we find that the 't Hooft-Polyakov monopole satisfies the well-known Dirac condition $\mu = 1/2e_{min}$. As for the mass of the SU(2) monopole, it can be shown that it must be of the same order of magnitude as m_v/α, where m_v is the mass of charged vector particles.

Regardless of the gauge in which we write the monopole, the vacuum is "spoiled" and does not coincide with the ordinary vacuum, even infinitely far from the monopole. In the case of the uncombed hedgehog it differs in being "uncombed," and, in the case of the combed hedgehog, it differs in that the Dirac string goes to infinity. This is the principal distinction of an isolated magnetic monopole from any other isolated ordinary particle. The whole universe has to be modified in order to create a single monopole.

If there is a monopole-antimonopole pair, the Dirac string ties them together and the vacuum at infinity remains ordinary. But this pair cannot exist for an indefinitely long time because, being attracted to each other, the monopole and the antimonopole will approach and ultimately annihilate.

Non-Abelian monopoles with a mass of about m_W/α do not appear in the electroweak theory because the group $SU(2) \times U(1)$ is neither simple nor semisimple. It contains an Abelian invariant subgroup. However, non-Abelian monopoles are predicted by grand unification models, and, in particular, by the $SU(5)$ model. The masses of such grand monopoles must be of the order of m_X/α_{GU}, that is, approximately 10^{16} GeV. Some grand monopoles must be colorless (purely electromagnetic) and some must be colored, possessing both magnetic and chromomagnetic charges.

Tremendously large masses of grand monopoles preclude their creation in accelerator experiments. What remains is to hunt primordial monopoles left over (?) as a remnant of the Big Bang. But how should one go about searching for them?

Relativistic monopoles must be highly ionizing because of their large magnetic charge. The unsuccessful search for dense ionization tracks imposes an upper bound on the flux of fast monopoles:

$$(10^{-4}-10^{-3}) \text{ m}^{-2} \text{ sterad}^{-1} \text{ day}^{-1}.$$

Slow monopoles ($v/c < 10^{-4}$) do not ionize atoms and cannot be detected by their ionization tracks. However, they can be detected by a current jump in a superconducting coil. A recent experiment at Stanford University with such a coil gave the upper bound

$$0.6 \text{ m}^{-2} \text{ sterad}^{-1} \text{ day}^{-1}$$

and one event that looked like a monopole traversing the coil.

A new topic became a subject of active discussion in 1982. According to the line of argument suggested by Rubakov, processes of the type

$$p + monopole \rightarrow e^+ + monopole + mesons \text{ and/or a pair of leptons}$$

could have cross sections comparable to those of conventional hadronic inelastic processes. The monopole acts as a catalyst for proton decay: The monopole itself is unchanged and, in principle, can successively destroy an unlimited number of protons. If the induced decay cross section is large, a monopole traversing a detector designed to look for proton decays should produce a chain of practically simultaneous proton decays. This would be a very beautiful phenomenon.

It has even been suggested that the proton decay catalysis by trapped monopoles could be used for energy production. Optimists state that, in this way, mankind could achieve the final solution to the looming energy crisis.

It would hardly be justifiable to discuss seriously the prospect that power plants of the twenty-first century will use the energy of proton decay induced by a handful of magnetic monopoles (do not forget that, with an atomic weight of around 10^{16} GeV, one gram-atom of monopoles would weigh 10^{16}g $= 10^{10}$ tonnes, so that "a handful of monopoles" would be quite heavy). Much is not yet clear in estimates of the cross section of the induced decay process; we do not know whether the universe contains magnetic monopoles; and, finally, the very idea of grand unification and baryon nonconservation will be a daring theoretical hypothesis until "ordinary," spontaneous proton decay has been discovered. But even a skeptic will have to agree that the "thought experiment" on the utilization of monopoles is very instructive. It illustrates how the most abstract theoretical constructions may some day lead to a great technological revolution.

MODELS, MODELS, MODELS . . .

Among the groups used in the literature to construct grand unification models, the group SU(5) is special in that it has the lowest rank (namely, 4). Its advantage lies in the relative rigidity of its predictions. As for its shortcoming, we can mention that fermions of different generations are not unified and even fermions within a single generation are not described by a single irreducible representation of the group. This last shortcoming is absent in the orthogonal group SO(10), whose rank is 5 and which contains SU(5) as its subgroup: SO(10)⊃SU(5). The spinor multiplet of the group SO(10) contains 16 components and unifies all the left-handed spinors of one generation, including the left-handed antineutrino,

$$16_{SO(10)} = (1 + 5 + 10)_{SU(5)}$$

The group SO(10) allows the existence of certain phenomena forbidden in SU(5). Thus, in the framework of SU(5), nonconservation of the baryon quantum number B is rigidly related to nonconservation of the lepton quantum number L, so that the difference $B-L$ is conserved (as, for example, in the decay $p \to e^+\pi^\circ$). In SO(10), the conservation of $B-L$ can be violated if the set of Higgs fields is sufficiently rich. Experimentally, this can be manifested by neutron-antineutron oscillations in the vacuum.

From the standpoint of baryon number nonconservation, the transformation of a neutron into an antineutron is equivalent to the transformation of two neutrons into mesons and/or photons. In both cases, $|\Delta B|=2$. Vacuum oscillations n ⟷ ñ seem to be more interesting than decays of nuclei with $|\Delta B|=2$ because the process of oscillation is faster than the decay of nuclei. Indeed, the decay rate of two nucleons in a nucleus is proportional to the square of a matrix element with $|\Delta B|=2$, while the vacuum transition n ⟷ ñ is proportional to the first power of the matrix element with $|\Delta B|=2$. Since the energy release in the decay is $2m$, where m is the nucleon's mass, dimensional arguments show that

$$\frac{1}{\tau_{\text{decay}}} \simeq \left(\frac{1}{\tau_{\text{ocs}}}\right)^2 \frac{1}{m}.$$

Assuming that $\tau_{\text{decay}} \simeq 3 \times 10^{31}$ years $\simeq 10^{39}$ s and recalling that $1/m \simeq 10^{-24}$ s, we find that the expected period of the n ⟷ ñ oscillations is about 3×10^7 s, that is, about one year. In principle, vacuum transitions with such periods can be observed in intense beams of reactor neutrons. Obviously, it is impossible to watch a neutron flux for a year. Nor is it necessary. Transformations of single neutrons will occur in a sufficiently intense beam during a fraction of the first second of observation. The transformation can be recorded by putting a target in the way of the beam and observing the large release of energy due to the annihilation of antineutrons in the target. Such experiments are being prepared now in a number of nuclear reactor laboratories.

The higher the rank of a group, the richer are multiplets and the larger the number of hypothetical particles it contains. Thus, for example, the exceptional group E_6 of rank 6, for which SO(10) is a subgroup, contains fermions of one generation in a 27-plet,

$$(27)_{E_6} = (16 + 10 + 1)_{SO(10)}.$$

Among the eleven additional particles there are both new quarks and new leptons. The authors of such models assume that these additional particles have not been observed so far because of their superheavy masses.

The spontaneous breaking in the minimal SU(5) model proceeds in two stages. The first stage is characterized by a scale 10^{14}–10^{15} GeV; here SU(5) breaks down to $SU(3) \times SU(2) \times U(1)$. The second stage is characterized by a scale 10^2–10^3 GeV. Here the symmetry is broken down to $SU(3)_c \times U(1)_{em}$. These two scales are separated by a dull gauge desert without either new particles or new physical phenomena. In higher-rank groups, the desert is interspersed with oases. As we go downward in the energy scale, another intermediate symmetry is broken in each such oasis. If the original group is sufficiently complex, it may allow for different alternative versions of oases. For instance, the first stage in the breaking of the group SO(10) can be either the group SU(5) or a left-right-symmetric group $SU(4) \times SU(2)_L \times SU(2)_R$. It should be noted in passing that the first-ever version of grand unification was suggested by Pati and Salam in 1973, precisely on the basis of the group $SU(4) \times SU(2)_L \times SU(2)_R$.

Among the theoretical models operating with high-rank groups, those in which unification covers fermions not only within a single generation, but also in different generations are of special interest. Several types of such "truly grand" models have been discussed:
(i) The orthogonal groups SO(18), SO(22) . . . ;
(ii) The unitary groups SU(8), SU(14) . . . ;
(iii) The exceptional groups E_6, E_7, E_8;
(iv) The products of simple groups related by a discrete symmetry, so that there is only one gauge constant: $SU(5)^2 \equiv SU(5) \times SU(5)$, $SO(10)^2$. . . .

These models contain horizontal-symmetry gauge bosons giving the so-called horizontal transitions between generations,

$$\nu_e \longleftrightarrow \nu_\mu \longleftrightarrow \nu_\tau,$$
$$e \longleftrightarrow \mu \longleftrightarrow \tau,$$
$$u \longleftrightarrow c \longleftrightarrow t,$$
$$d \longleftrightarrow s \longleftrightarrow b.$$

If the masses of "horizontal" bosons are not too large, then there must exist rare decays of the type $\mu \to e + \gamma$. It is of special interest to look for such decays. The search can be conducted at meson factories, i.e. large-current proton accelerators with energies less than or of the order of 1 GeV.

The senior of the exceptional groups, E_8, is a singular point in the ocean of models. Its interesting property is that the dimensions of the fundamental and adjoint representations are identical: the E_8 model contains 248 fermions and 248 gauge bosons. Unfortunately, the number of scalar bosons in this model runs into the thousands. This feature scares theorists off, and no detailed analysis of the E_8 model can be found in the literature.

One of the most serious shortcomings of all the models mentioned in this section is that they do not solve the so-called problem of mass hierarchy, that is, the problem of the hierarchy of energy scales. The hierarchy problem looks simplest in the SU(5) model containing only two scales: 10^{14} and 10^2 GeV. Formally, we can write a Higgs potential that has two independent energy minima: one at high energy and another at low energy. However, it is difficult to isolate them physically from each other. The heavy and light Higgs bosons interact, because of quantum effects, through the exchange of gauge bosons. Hence, even if we achieve "hyperfine tuning" of the difference between vacuum condensates in the original Lagrangian by 12 orders of magnitude, this tuning will be destroyed by quantum corrections and the mass of the W bosons will grow to about the mass of the X bosons.

Another manifestation of the same problem can be noticed even when ignoring the grand unification and remaining in the framework of the standard SU(3)×SU(2)×U(1) theory. The point is that diagrams contributing to the mass of Higgs bosons diverge quadratically (see Figure 34, in which wavy lines represent spin 0, smooth lines spin 1/2, and dashed lines spin 1 particles). If these diagrams do not mutually compensate one another, the cutoff of quadratic divergences will not be realized below the Planck mass. This would mean that the natural mass scale of the Higgs bosons, and consequently of the W bosons, has to be of the order of the Planck mass. In other words, this would mean that the Fermi constant G_F must be of the order of the Newton constant G_N, whereas, in nature, the former is 33 orders of magnitude greater than the latter.

This hierarchy paradox could be solved if the divergences of the diagrams in Figure 34 mutually cancelled out. In principle, there is some

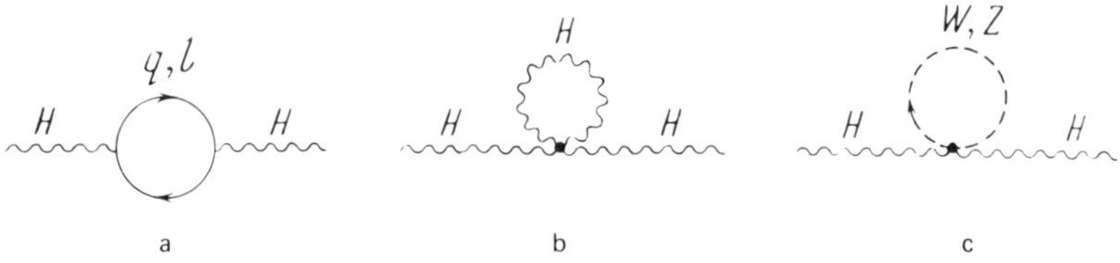

FIGURE 34. The contributions of (a) quarks and leptons, (b) Higgs bosons, and (c) intermediate bosons to the mass of Higgs bosons.

possibility of that because the fermion loop has a minus sign and boson loops enter with a plus sign. But such a cancelling-out calls for a symmetry between fermions and bosons.

Again, as we witnessed more than once in the theory of elementary particles, it occurred that such a symmetry was already being studied by theorists driven by a pure scientific curiosity, with no connection to the problem of scale hierarchy. Moreover, this symmetry already had a name: supersymmetry.

SUPERSYMMETRY

Despite the salient differences between such symmetries as isotopic, color, electroweak and grand unification symmetries, all these have an essential common feature: they are all internal symmetries. The word "internal" signifies that transformations of these symmetries do not affect the space-time geometric properties of the transformed states. Isotopic rotations can turn a neutron into a proton, a particle with the same spin, but cannot turn a neutron into, say, a π meson. Transformations under the SU(5) symmetry transform leptons and quarks into one another, but again these are states with a given spin (and a given helicity).

In contrast to purely internal symmetries, transformations of supersymmetry transform fermions and bosons into one another; for instance, they transform a scalar into a spinor particle, or a spinor into a vector particle. The first paper on supersymmetry was sent by Gol'fand and Likhtman to *JETP Letters* in 1971; but even today, at the end of 1982, when I write these lines and the number of papers on supersymmetry reaches into the thousands, we discern no traces of supersymmetry in the spectrum of known elementary particles. Why then does supersymmetry (specialists abbreviate it to SUSY) enjoy such popularity? What caused this "superrush"? Skeptics reply: vogue. Enthusiasts reply: grand expectations.

Supersymmetry indeed gives rise to superexpectations. Consider a supermultiplet, for example, one of the simplest supermultiplets, containing only two particles, the photon and the photino; that is, a hypothetical, truly neutral spin 1/2 particle resembling the Majorana neutrino. Transformations within a supermultiplet are realized by a spinor generator Q. The nontriviality of these transformations is clear from the fact that fields in this supermultiplet have different dimensions: m for the boson field and $m^{3/2}$ for the fermion field. This is reflected in the expression for the anticommutator of the two spinor generators Q: it is expressed in terms of a quantity of dimension m, namely through the 4-momentum p_μ (the generator of four-dimensional translations),

$$\{Q,\bar{Q}\} \equiv Q\bar{Q} + \bar{Q}Q = -2p_\mu\gamma_\mu,$$

where γ_μ are four Dirac matrices. The spinor transformation resembles the square root of a translation.

We again see that the spin is not an internal variable. This conclusion is not new to us; for instance, the spin enters as an equal partner, together with the orbital angular momentum, into the expression for the total angular momentum.

Generators of the space-time translations p_μ and rotations $M_{\mu\nu}$ form, together with spinor generators Q, the so-called graded algebra of supersymmetry, which contains the Poincaré algebra as a subalgebra (an algebra is said to be graded if it comprises anticommutators together with commutators). Supersymmetry thus offers a generalization of the Poincaré group and an elaboration of special relativity. Geometric translations and rotations do not affect the nature of a particle: the electron remains itself under all translations and rotations. The internal transformations that we have studied up to now do not alter the coordinates of a particle. But now we find that, having transformed one particle into another and then gone back to the original one, we discover it in a different point of space.

It should be remarked that manifestations of a relationship between spatial and internal variables have already been encountered. What I mean here is the CPT theorem. A relation between the charge conjugation C and time reversal T is built into the foundations of quantum field theory. Recall that the same operator creates a particle and annihilates an antiparticle and that, in the language of Feynman diagrams, a positron is an electron moving against the time arrow. As for the P transformation, it is involved because of the isotropy of the four-dimensional Euclidean space.

The isotopic and spatial coordinates are related in a nontrivial manner in non-Abelian monopoles. But this is a property of a spatially extended object, a property of the solution, not of the Lagrangian of the theory.

Supersymmetry discovers new profound relationships between different types of transformations in the mathematical apparatus of the theory. In going from a global supersymmetry (Volkov, Akulov, 1972; Zumino, Wess, 1974) to a local supersymmetry in which the parameters of transformation are functions of space-time coordinates, we obtain a generalization of general relativity: supergravity (Deser, Zumino, Ferrara, Freeman, and von Nieuwenhuisen, 1976). The simplest supergravity multiplet contains two particles: a graviton and a gravitino, the hypothetical spin 3/2 neutral particle.

Fermions and bosons are present symmetrically in the same multiplet; this holds the promise of cancelling undesirable divergences both in the global supersymmetry and in supergravity because, as we have already mentioned, fermion loops have a minus sign and boson loops a plus sign.

A quantity of enormous interest, for which the discussed compensations may prove vitally important, is the so-called cosmological constant λ, that describes the gravitational "charge" of the vacuum, that is, the gravitational density of the energy-momentum tensor of the vacuum. No effects of the cosmological constant have ever been detected in experiments and λ is usually set equal to zero. Astronomical observational data indicate that $\lambda < 10^{-47}$ GeV^4 (this limit approximately corresponds to one proton mass per cubic meter of vacuum).

On the other hand, naive dimensional estimates of the vacuum fluctuations of any given field make us anticipate $\lambda \sim m_p^4 \sim 10^{78}$ GeV^4 where m_P is the Planck mass. Could this mean that nature realizes superhyperfine compensations of contributions due to different fields? Unfortunately, it is not clear at the present moment how such fantastically accurate compensations could be realized not in the strict, but in the broken supersymmetry. Indeed, even if SUSY is used by nature at all, it is very strongly broken. In fact, we have not even seen a single supermultiplet in our experiments, and must hope that superpartners of our ordinary particles have not been discovered because their masses are much larger.

We have already mentioned that supersymmetry offers a unique opportunity to unify internal and geometric symmetries. This unification is technically realized by "labelling" the spinor generator Q with some internal index i ($1 \leq i \leq N$). The generator Q_i changes not only the spin but also the "flavor" of particles. The corresponding theoretical structure is called extended supersymmetry.

A special example of extended global supersymmetry is the case of the index i running through integers from 1 to 4. This is the so-called $N=4$ supersymmetry. This theory comprises 11 massless particles: one with $J=1$, four with $J=1/2$, and six with $J=0$, comprising eight bosonic and eight fermionic helicity states. (I suggest that the reader reconstruct all these numbers himself, starting with the right-handed helicity state of a vector particle and taking into account the fact that the generator Q reduces the spin of a state by 1/2 and that $Q_iQ_j\psi=0$ if $i=j$.) Of special interest is the model which has an arbitrary gauge symmetry and includes an $N=4$ global SUSY as an external factor. An example of such a model is a structure with gauge symmetry SU(2) and an isotopic triplet of gauge fields in which each of the three particles in the triplet is included in its own 11-component supermultiplet. It has been recently found that the running gauge constant in such a model stops running. Direct calculations show that the gauge constant is independent of momentum in the one-, two-, and three-loop approximation. If this effect holds in all orders of perturbation theory, this theory is conformally invariant and finite for all momenta.

Extended supergravity opens even more breathtaking prospects. Here a unification of internal and geometric degrees of freedom offers the hope of superunifying all fundamental forces of nature, including the gravitational force.

By consecutively applying spinor generators Q_i, we can readily show that the maximally extended supergravity not containing particles with $J > 2$ corresponds to $N=8$. In this case, the supermultiplet contains the following massless particles: one graviton, eight gravitinos, 28 bosons with $J=1$, 56 fermions with $J=1/2$, and 70 scalar particles (128 bosonic and 128 fermionic helicity states in all). This theory has a global SO(8) symmetry. The $N=8$ supergravity is singled out from other supergravity theories with lower values of N in the same way that the $N=4$ supersymmetry is from other global supersymmetries. The absence of ultraviolet divergences has been proved for the $N=8$ supergravity for the maximum number of loops. Its behavior at short distances is much less singular than that of ordinary gravitation and possibly less singular than that of its "junior sisters" with $N<8$.

Interesting attempts were made to construct, on the basis of the $N=8$ supergravity, a model that would have local SU(8) symmetry and contain both the group SU(5) as one of its subgroups for each of the three generations and a horizontal symmetry group relating different generations. The distance covered along this road is still much shorter than that still separating us from the goal.

A large number of papers has been published during the past two years on the introduction of the $N=1$ supersymmetry (first global, then local) into SU(5)- or SO(10)-type models of grand unification. This problem is not as stupendous as that of superunification. Here the authors formulate a more modest objective: to solve the problem of the mass hierarchy mentioned at the end of the last section.

The price that has to be paid for cancelling divergences is the doubling of the number of all known fundamental particles: each particle must have a superpartner. Some of these superpartners have generally accepted names: the photino and the gluino, but the names for other particles are not yet definitely established. Scalar superpartners of leptons are often called sleptons (e.g., selectron), spinor partners of Higgs bosons are called shiggses, and those of hadrons are called shadrons. These terms lack somewhat in euphony; in my opinion, it would be more convenient to form the names for all superparticles in a unified manner: by a suffix ino; for example, electrino, nuino, muino, higgsino, hadrino, denoted by the primed symbols of the appropriate particles: e', ν', μ'. . . .

One could object that the diminutive Italian suffix ino ("neutrino" means a little neutron) hardly suits superparticles that are certainly much heavier than their ordinary kinfolk. But physics terminology retains quite a few such cosy historically traceable oxymorons that we habitually overlook: atoms are divisible, the proton is not that simple, and some mesons are much heavier than many a baryon. But this is merely an aside remark. The problem is certainly not what to call superparticles but how to predict their properties, especially their masses.

Obviously, to produce the cancellation in the masses of the scalars that we discussed earlier, the masses of inos must be substantially less than 1 TeV. Otherwise, the Higgs vacuum expectation value in the electroweak theory would be much greater than its known value (about 200 GeV). In order to predict the masses of superparticles with better accuracy, we have to choose one of the numerous concrete models with a specific set of particles and a specific mechanism of supersymmetry breaking.

A phenomenological analysis of the available experimental data shows that the photinos may be just as light as ordinary neutrinos. On the other hand, the lower bound for the mass of the gluino and of the superhadrons containing it (hadrino) comes to several GeV. A search for such hadrinos is possible on existing proton accelerators.

The existence of superparticles must affect the proton lifetime because the rate of running of the running constants (the slopes of the trajectories in Figure 32) is a function of the number and type of supermultiplets; hence, the grand unification mass will be also affected. Furthermore, this gives rise to new mechanisms of proton decay.

Even if all superparticles are heavy and their masses are in the neighborhood of 100 GeV, the discovery and analysis of at least some of these particles appears to be a quite feasible problem for the next generator of colliders and could, under favorable conditions, be realized during the next decade.

The discovery of superparticles would constitute a triumph of symmetry ideas. Besides, it would definitely tell us a lot about the mechanisms of symmetry breaking.

UNIFICATION MODELS AND THE BIG BANG

The truly astronomical numbers characterizing the Planck mass m_P and the grand unification mass m_{GU} in our usual units kill all hopes that we shall ever be able to experiment with such energies on accelerators.

There exist quite realistic projects for a very large proton collider with a proton energy in the tens of thousands of GeV. New acceleration techniques may ultimately bring the energy of protons up to more than 10^5 GeV. But even a fantastic accelerator constructed as a superconducting ring, free-floating in space, with a diameter greater than that of the earth, could accelerate protons only up to 10^8 GeV, which is still much less than m_{GU}. (The limit 10^8 GeV is imposed by the synchrotron radiation: the energy emitted per turn increases as the fourth power of the energy of the particle for a fixed orbit radius).

Experimental physics in the vicinity of m_{GU} could be possible if we managed to catch and slow down grand monopoles. Indeed, the annihilation of a monopole and an antimonopole must create the X and Y bosons and very heavy Higgs bosons characteristic of grand unification schemes. However, if we recall the high hopes related to grand monopoles as catalysts of proton decay, it can be anticipated that monopole shooting licenses, if and when such are offered, would be very hard to come by.

The lack of earth-bound prospects makes theoretical physicists, occupied with unification problems, turn to cosmology and, specifically, to the first moments of the Big Bang.

We know that the hot universe theory gives the following relation between the age t and temperature T of the universe,

$$t \sim m_P/T^2,$$

or, in the usual units,

$$t \text{ (seconds)} \sim 1/T^2 \text{ (MeV)}$$

A temperature of the order of 10^{15}–10^{19} GeV thus corresponds to an age of 10^{-36}–10^{-44} s. Such a superyoung universe is a natural laboratory for testing grand unification and superunification models. Unfortunately, no "eyewitnesses" of these first moments survived, but certain indirect reconstructions are possible from some properties of the surrounding universe.

Among such fundamental properties of the world are the age of the universe, $(1-2)\times 10^{10}$ years; the Hubble law of the recession of galaxies; the existence of the primordial photon gas (microwave background radiation) at a temperature $T \simeq 3$ K; the homogeneity and isotropy of that gas; the mean density of visible matter, which corresponds to approximately one proton per cubic meter; and, finally, a comparable, or even greater, density of invisible matter in coronas of galaxies and of galactic clusters.

One of the key parameters of the universe is the ratio of the number of nucleons in the universe to the number of background photons. Observations show that this ratio is 10^{-10}–10^{-9}. In 1967, A.D. Sakharov suggested that this ratio is a corollary of a tiny $(1 + 10^{-9})$:1 excess of nucleons over antinucleons that was formed in the early universe due to the nonconservation of baryonic charge and CP violation. Both these components are present in the models of grand unification developed later, in the 1970s. The third component necessary for the survival of baryonic excess is the nonequilibrium due to the expansion of the universe.

A copious literature is available on the calculations of the baryonic assymetry of the universe in different models of grand unification and superunification. These calculations show the important role played in the formation of the baryonic excess not only by the X and Y bosons but also by superheavy Higgs bosons. The results of calculations depend especially strongly on the detailed (and so far unknown) properties of these bosons. These properties also determine the dynamics of the cooling of the universe. As a result, it is not yet possible to calculate baryonic asymmetry unambiguously. However, cosmologists insist that they are able to calculate baryonic asymmetry "to the end" in the framework of any concrete scheme of grand unification. Hence, in the future, the ability to explain the baryonic asymmetry of the universe quantitatively will be one of the central criteria for selecting the winner in the beauty contest of grand models.

With the baryonic charge nonconserved and CP parity not violated, practically no matter except photons and neutrinos would survive in the hot universe. Obviously, everything surrounding us and we ourselves owe our existence to the very weak violation of CP invariance. Just imagine that, until 1964, most physicists believed that CP was conserved!

Another important statement preceding the work on grand- and supercosmology was made in 1972 by Kirzhnits and Linde, who noticed that spontaneously broken symmetries must be restored at a sufficiently high temperature. Consequently, during the first moments of cooling, the universe had to go through stages of successive breaking of symmetries. The number of stages is only two in the minimal SU(5) model: breaking of SU(5) to SU(3) × SU(2) × U(1) at 10^{14} GeV and to SU(3)×U(1) at 10^2 GeV. The number of such stages is greater in more complicated models.

The restoration of symmetry at sufficiently high temperatures is illustrated in Figure 35. Figure 35a shows the Higgs potential at zero temperature (here φ_0 is the vacuum expectation of the Higgs field; compare it to Figure 31). Figure 35c shows the effective potential at a very high temperature above the phase transition temperature $(T \gg \varphi_0)$. Figure 35b corresponds to an intermediate temperature.

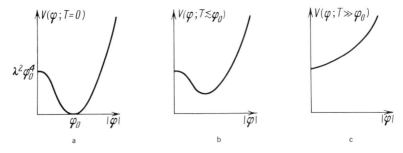

FIGURE 35. The effective potential $V(\varphi,T)$ for three different temperatures.

We see from Figure 35c that no vacuum scalar condensate exists at high temperatures. It appears when the universe cools down. If the cooling process is sufficiently rapid, so that no exchange of signals occurs between different regions of space, phases of the condensate are formed in these regions independently and are not correlated (see Figures 36 and 37). And

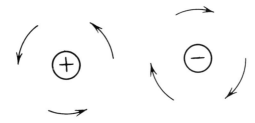

FIGURE 36. The cross sections of tubes resulting from the spontaneous breaking of the Abelian U(1) symmetry. The fluxes in the tubes run counter to each other.

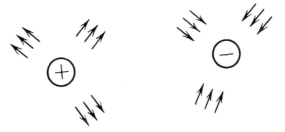

FIGURE 37. A monopole-antimonopole pair resulting from the spontaneous breaking of a non-Abelian symmetry.

if the Abelian symmetry U(1) is thereby broken, strings, or tubes, are formed in the vacuum, with the flux of the gauge Abelian field being squeezed into these strings by the surrounding Higgs condensate. Figure 36 shows a cross section by a plane of two such tubes in which the fluxes are in opposite directions. But if a non-Abelian gauge symmetry is broken, then hedgehog monopoles appear in a similar manner in the cooling primordial jelly (see Figure 37, where the formation of a monopole-antimonopole pair is shown, and compare it to Figure 33).

If a discrete symmetry is spontaneously broken, then the condensate of the matter field φ in the neighboring regions of space may have opposite signs ($<\varphi> = \pm\varphi_0$, see Figure 38). The boundaries between vacuum

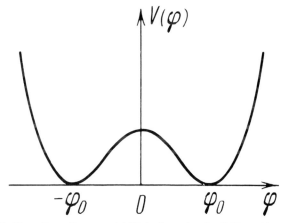

FIGURE 38. The self-coupling potential $V(\varphi)$ of a real scalar field φ with two degenerate minima, at $\varphi = \varphi_0$ and $\varphi = -\varphi_0$, corresponding to two degenerate vacua.

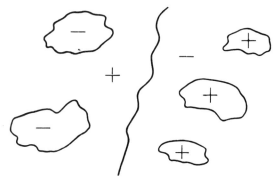

FIGURE 39. Vacuum domains with $\varphi = +\varphi_0$ and $\varphi = -\varphi_0$. The domains are separated by very thin but very massive walls.

domains with positive and negative condensates emerge as very thin and very dense walls: The wall thickness is of the order of $1/\lambda\varphi_0$ and the surface density is of the order of $\lambda\varphi_0^3$, where λ^2 is a dimensionless constant of nonlinear self-interaction of the Higgs field. A plane section of such a vacuum domain is shown in Figure 39.

The Cosmological creation of strings and monopoles was first discussed in 1976 by Kibble. The formation of vacuum domains, the properties of domain walls, and their effect on the evolution of the universe were first analyzed by Kobzarev, Zel'dovich, and myself in 1974, on the basis of a model Lagrangian proposed by Lee and Wick. At the same time, Kobzarev, Voloshin, and I discussed the problem of the decay of a metaestable vacuum.

A vacuum is metastable if the minimum of potential energy corresponding to it is not the absolute minimum. In Figure 40, the metastable vacuum

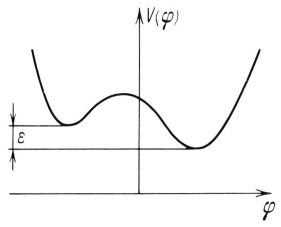

FIGURE 40. The self-coupling potential $V(\varphi)$ of a real scalar field φ with two nondegenerate minima. The upper minimum corresponds to the so-called false (unstable) vacuum. The lower minimum corresponds to the stable vacuum.

corresponds to the minimum on the left and the stable vacuum to the minimum on the right. A transition from a metastable vacuum to a stable vacuum releases energy, but Figure 40 shows that this should be a barrier-penetration process, forbidden in classical physics. The transition proceeds by the quantum creation of a bubble of new vacuum, which then classically expands, the rate of this expansion rather rapidly approaching the velocity of light.

The smaller the volume of a newborn bubble, the higher the probability of its creation. The minimal, or critical, bubble radius R_c obtains from energy arguments: The gain in energy in the volume of the bubble must balance out the energy loss due to the surface energy of the walls,

$$\frac{4\pi}{3} R_c^3 \epsilon = 4\pi R_c^2 \sigma;$$

hence, $R_c = 3\sigma/\epsilon$. Here ϵ is the difference in energy density for two vacua (see Figure 40) and σ is the surface density of the wall.

It is easy to show that the probability of the formation of a bubble with radius R_c via barrier penetration is determined by the exponent

$$\exp\left(-\frac{\pi}{2}\sigma R_c^3\right)$$

and may be vanishingly small for reasonable values of σ and ϵ ($\sim 10^{-100}$–10^{-1000}). (The pre-exponential factor has so far not been calculated, despite the plentiful literature devoted to vacuum bubbles.)

The theory of vacuum bubbles contains a number of very interesting theorems. Thus, Coleman noticed that an expanding bubble is invariant under Lorentz transformations (the center of the bubble is at rest in an arbitrary inertial frame of reference). Coleman also found that, in some cases, the upper vacuum does not decay at all because of the stabilizing effect of gravitation.

In order to illustrate the potential role of two- or multistory vacua in cosmology, let us consider a simple equation describing the expansion of the universe,

$$\frac{1}{2}\dot{a}^2 - \frac{4\pi}{3}G_N \rho a^2 = K,$$

where a is the distance between any two points (a is often called the scale factor), the dot over a denotes the derivative with respect to time, G_N is the Newton constant, ρ is the mean energy density in the universe, and K is a constant. In effect, the first term is the kinetic energy of a unit-mass probe particle and the second term is its potential energy; their sum K does not change in the process of the evolution of the universe.

The sign of the constant K describes the type of evolution. If $K < 0$, the universe is closed and its expansion will sooner or later be replaced by contraction. If $K > 0$, the universe is open and will expand indefinitely. The boundary mode $K = 0$ is called the flat universe mode. In this case, the spatial curvature of the universe is zero and the three-dimensional space is Cartesian. All the available knowledge about our universe does not contradict the assumption that the universe is flat (Cartesian).

The expansion equation for the universe is usually written in the form

$$\left(\frac{\dot{a}}{a}\right)^2 \frac{8\pi}{3} G_N \rho = \frac{2K}{a^2}.$$

The quantity $H = \dot{a}/a$ is called the Hubble constant. The current value of the Hubble constant is obtained by measuring the velocity of receding galaxies, $\dot{a}$, and the distances to them, a, and, according to one set of data, approximately equals 50 km s^{-1} megaparsec^{-1}, while, according to another set, it is twice as large.

The value of H changes, of course, as the universe gets older. The evolution $H(t)$ is determined by the dependence of the mean energy density in the universe, ρ, on time.

At present, the value of ρ is mostly determined by the masses of nonrelativistic particles: these particles behave like cold dust. Obviously, in this case $\rho \sim a^{-3}$.

At the earlier stages, when the universe was sufficiently hot, it was dominated by a gas of relativistic particles ("photon gas"). Simple dimensional arguments show that, in this case, $\rho \sim a^{-4}$. (The density of photons diminishes as a^{-3}, while the energy of each photon diminishes as a^{-1} because the photon wavelength increases with increasing scale factor. Compare this to the familiar relationship $\rho \sim T^4$ for blackbody radiation.)

It will be easy to understand that $\rho \sim a^{-2}$ for a universe filled with strings, $\rho \sim a^{-1}$ for a universe filled with walls, and, finally, $\rho \sim a^0 =$ const for a universe with a nonzero cosmological term (the universe with, so to speak, massive vacuum). By solving a simple differential equation, we can easily see that $a \sim t^{1/2}$ ($t^{2/3}$, t, t^2, and e^t) for a universe filled with photons (dust, strings, walls, and massive vacuum, respectively). We shall address the exponential expansion of the universe, $a \sim e^t$, with a nonzero cosmological term below, but first we take up a very interesting aspect related to the constant K.

The density ρ corresponding to a flat universe (the case $K = 0$) is usually called the critical density and is denoted by ρ_c. Obviously,

$$\rho_c = \frac{3H^2}{8\pi G_N}.$$

The ratio of the observed density ρ to the critical density ρ_c is denoted by Ω: ($\Omega = \rho/\rho_c$). If $\Omega > 1$, the universe is closed; if $\Omega < 1$, the universe is open; and if $\Omega = 1$, the universe is flat.

There is no unanimous opinion in the literature on the value of Ω but I would say that most authors agree that $0.1 \lesssim \Omega \lesssim 2$. (Note that $\Omega_B =$

$\rho_B/\rho_c \lesssim 0.03$, where ρ_B is the density of baryons. The rest comes from the invisible matter.)

The fact that by the order of magnitude today's Ω is close to unity means that the quantity $\Omega - 1$ was unimaginably close to zero at the onset of the Friedmann expansion. Indeed, the definition of ρ_c yields $\rho_c - \rho = 3K/4\pi G_N a^2$, where K is a constant independent of time. On the other hand, in a gas of relativistic particles, $\rho \sim a^{-4}$; therefore,

$$(\Omega - 1)_{a=a_1} / (\Omega - 1)_{a=a_2} = a_1^2/a_2^2$$

(here for simplicity's sake we choose to neglect that the last stage of the evolution of the universe is dominated by dust with $\rho \sim a^{-3}$ and not by relativistic gas with $\rho \sim a^{-4}$). If we take the relation $a_1/a_2 = T_2/T_1$ into account and take the current temperature $3\,K$ for T_2 and the grand unification temperature $T_{GU} \simeq 3 \times 10^{14}$ GeV for T_1, we obtain

$$(\Omega - 1)_{T=3 \times 10^{14}\,\text{GeV}} : (\Omega - 1)_{T=3K} \simeq \left(\frac{3 \times 10^{-4} \times 10^{-9}}{3 \times 10^{14}}\right)^2 = 10^{-54}.$$

Then, from $(\Omega - 1)_{T=3K} \sim 1$, it follows that $(\Omega - 1)_{T_{GU}} \simeq 10^{-54}$. What factors could cause this fantastic accuracy in the mutual compensation of H^2 and $\frac{8\pi}{3} G_N \rho$ at a temperature of about T_{GU}?

Let us try to answer this question by starting to watch the universe at a moment $t \sim t_P \sim 1/m_P$ and assuming that, at this moment, the difference $2K/a^2 = H^2 - \frac{8\pi}{3} G_N \rho$ was of the order of m_P^2 in magnitude. Were this difference negative ($K < 0$), the universe would oscillate with the Planck period $1/m_P$. Were it positive ($K > 0$), the universe would expand over a time $1/m_{GU}$ and cool down to a temperature $T_{GU} \sim m_{GU}$ (we emphasize that this would occur over a time $1/m_{GU}$ and not over the much longer time m_P/m_{GU}^2 characteristic of a Friedmann expansion of the photon gas).

The case we discuss below is that of $K > 0$. As the linear expansion due to the K-term continues ($a \sim t$), the contribution of the relativistic gas, $\sim G_N a^{-4}$, becomes progressively less significant, but the contribution of the cosmological term $G_N^2 \lambda^2 \varphi^2$ becomes dominant at $T \sim m_{GU}^2/m_P$. Here λ^2 is the dimensionless self-coupling constant of the Higgs field and φ_0 is the Higgs expectation value (see Figure 35a), $\varphi_0 \sim m_{GU}$. The cosmological term arises because the mean Higgs field, which equals zero at $T \sim m_P$, did not have enough time to reach its condensate value φ_0 (see Figure 35).

The linear expansion of the universe changes, owing to the cosmological term, to an exponential expansion

$$a \sim e^{(m_{GU}^2/m_P)t}.$$

In this so-called de Sitter mode, the contribution of the K-term dies out exponentially with time (as a^{-2}), and it is sufficient for $t \sim 70(m_P/m_{GU}{}^2) \sim 10^{-33}$ s to elapse in order for the term $2K/a^2$ to become 54 orders of magnitude less than each of the terms H^2 and $\frac{8\pi}{3} G_N \rho$. Hence, the exponential inflation of the universe purges it of the K-term, but this produces an empty and supercooled universe.

The theory of the inflationary universe was born in 1980. A large number of papers has been devoted to it in recent years, with different specific scenarios differing from one another in the mechanisms of the creation and destruction of the cosmological term. Some authors (Starobinsky) have the cosmological term created by quantum gravitational effects, others (Guth, Hawking) derive it from a standard Higgs potential (see Figure 35), and still others (Linde) from the Coleman-Weinberg Higgs potential discussed in Chapter V (in this last case, the effective potential $V(\varphi,T)$ contains a metastable minimum at $\varphi = 0$ at a nonzero temperature; see Figure 41).

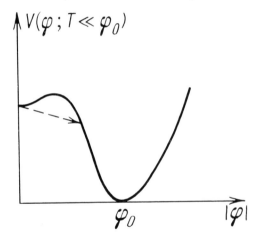

FIGURE 41. The potential V(φ,T) with a metastable minimum at $\varphi=0$ and a stable minimum at $\varphi=\varphi_0$. The arrow marks the tunneling formation of the bubble of stable vacuum.

In all Higgs scenarios, the metastable vacuum with a nonzero cosmological term decays through the formation of bubbles of a new vacuum. In the scenario suggested by Linde, this is a single bubble whose size at present exceeds, and by many orders of magnitude, the size of the whole visible universe. The approximate tunneling path of this bubble is shown by the arrow in Figure 41. You will notice that the bubble is created not empty but filled with a field ($\varphi - \varphi_0$) that ultimately transforms into a relativistic gas of particles whose temperature is less than the grand unification temperature by approximately an order of magnitude. This scenario explains the long life and Cartesian nature of the universe and predicts the absence of grand monopoles because the Friedmann expansion now starts at an insufficiently high temperature.

It goes without saying that such cosmological scenarios must be considered only as very rough preliminary sketches of answers to the question of how the initial conditions that determined the whole further evolution of the universe had been created. But it is extraordinary in itself that such questions can be discussed at all in a constructive manner.

ON EXTRAPOLATIONS AND PREDICTIONS

"Given that the theory has been so fantastically modified during the past 25 years, who can guarantee that, in another 25 years, it will not change so much that all current theoretical constructions will have been thrown on the dump?"

This is roughly the question often posed by skeptical listeners both after public lectures and during discussions at home in physicists' families.

As far as the experimental and theoretical discoveries of the past 25 years are concerned, I want to emphasize first of all that all of them are in the mainstream of the Lagrangian quantum field theory developed in the first half of this century. Although such objects as quarks and Higgs bosons are indeed very unusual, the theories describing them do not violate any of the principles established earlier. Niels Bohr could hardly classify them as the "crazy" theories he felt the need for at the end of the 1950s.

Looking back not only at the past 25 years but also at a longer period, one cannot but notice that, with each substantial change in the physical picture, new theories did not destroy their predecessors. Having a much wider range of applicability, they revealed the limits within which the older theories were valid and enveloped them as legitimate limiting cases. This is precisely what has happened in the transitions from nonrelativistic to relativistic mechanics, from classical to quantum mechanics, and from quantum mechanics to quantum field theory. The theory of the electroweak interaction did not eliminate but rather deepened and substantiated the theory of four-fermion coupling. The same can be anticipated for future theories.

These future theories will extend and refine our knowledge. They will describe new phenomena and will enable us to describe the known ones to a greater accuracy. But what we know we do know.

The modern theories are definitely reliable in the range of their applicability. But, in fact, when using these theories, we often extrapolate them very far, to regions where they have never been experimentally tested. Sometimes this extrapolation is made unconsciously, and contradictions and paradoxes are bound to appear when the extrapolation is hurt by facts. In most cases, though, the extrapolation is made quite consciously.

A brilliant example of a far-reaching extrapolation is provided by gravitation. I have already mentioned that, experimentally, the gravitational interaction has been measured only at distances not shorter than a few centimeters, while the classical theory of the gravitational interaction (general relativity) is assumed to be valid down to nearly Planckian distances, that is, of the order of 10^{-33} cm, where strong quantum corrections come into play.

Is there a foundation for such extrapolation? Yes, and beyond doubt. General relativity is a very beautiful theory; it would be rather flippant to discard or modify it without really compelling reasons, all the more so because there are no elegant theoretical models in which the gravitational interaction is modified at some intermediate distance. Physicists prefer to cut off such optional speculative modifications by Occam's razor.

I am nonetheless profoundly convinced that we must use every possibility to expand the domain of knowledge and contract the domain of faith. The principle of maximal possible extrapolation must be complemented by the principle of maximal possible verification. Thus, as far as gravitation is concerned, it would be of interest to verify Newton's law at the minimal possible distances. It would be very good to go down, even if to a poor accuracy, to a fraction of one millimeter. Note that the standard gravitational interaction has also not been verified quantitatively for large distances beyond the size of the solar system.

Gravitation is indeed a case in which extrapolation is made quite consciously, but, in some cases, as I mentioned earlier, this is not the case. For instance, there is a rather widespread opinion that the only existing long-range forces are the electromagnetic and gravitational. This indeed seems quite plausible. At any rate, no experimental basis exists now for introducing new long-range interactions. But it must be clearly understood that the experiments carried out up to now leave a rather large number of blank spots on the map of long-range interactions.

The better verified aspect is the absence of the so-called baryonic and leptonic photons, hypothetical massless vector particles whose sources would be the baryon and lepton charges, respectively. However paradoxical it may sound, the best upper bounds on the coupling constants α_B and

α_L of these photons are obtained from Eötvös's experiments, which were carried out at the beginning of this century, and some more accurate experiments of the same type carried out later.

The reader should know that Eötvös's experiments looked for a dependence of the pendulum oscillation period on the type of material from which the pendulum was made, that is, it tested, essentially, the equality of the inertial and gravitational masses. The gravitational interaction being very weak, even a small additional long-range interaction between the baryons and leptons of the earth and the pendulum would be manifested in these experiments as a departure from the equality of the inertial and gravitational masses. Indeed, the mass of the atom is not proportional to the number of electrons it contains and, owing to the nuclear mass defect, is only approximately proportional to the number of nucleons.

The upper bound on the coupling constant of baryonic photons, α_B, obtained from Eötvös-type experiments, is 10^{-47} and that on the coupling constant α_L of leptonic photons is 10^{-49}. This must be compared with the ordinary electromagnetic interaction for which $\alpha = 1/137$. It would appear, or at least, so it seems at the present time, that the existence of long-range interactions with constants α_B, α_L smaller than 10^{-50} is unrealistic. Bounds on other possible types of interactions are substantially cruder.

Thus, the emission and absorption of hypothetical neutral massless pseudoscalar particles could produce interesting spin effects resembling the interaction between the magnetic moments of particles and the electromagnetic field. However, in contrast to the usual electromagnetic field, the pseudoscalar field could not be screened out by metallic or any other screens.

The possibility of non-Abelian long-range interactions mediated by infinitely long and strong gluon-like strings cannot be excluded either. Since we do not presently know of any additional degeneracy of the known particles, these particles cannot act as quarks with respect to these new hypothetical gluons. But if such "quarks" exist and if they are not too heavy, then they are detectable in high-energy colliding-beam experiments. Any created "quark-antiquark" pair would then be tied with an extending string. However, it is time we returned from these unsubstantiated, although not refuted, fantasies to the mainstream of contemporary physics.

If we optimistically postulate that mankind will be able to emerge from the threatening military, economic, and ecological crises, retaining the basic human values in general and pure science in particular; if we further assume, just as optimistically, that the successful progress of elementary particle physics in the coming decades will not grind to a stop owing to various nonscientific factors; if, finally, we optimistically assume that the hopes of theoretical physicists will come true and the intermediate vector bosons,† scalar bosons, proton decay, neutrino masses, and superparticles will be discovered; and if we postulate that all these conditions are met, will it be possible to say then that the physical world will have been understood completely, that fundamental physics has reached the end of the road? It seems obvious to me that the answer must be in the negative.

The picture as outlined above contains too many fundamental particles, too many arbitrary parameters. In fact, the inner meaning of such fundamental concepts as spin and charge remains "a thing unto itself." In view of all this, it is natural to make the conjecture that there exists a deeper level of the physical world: subquarkian, subelectronic, and maybe even subphotonic.

The literature contains hundreds of speculative papers devoted to subparticles, and offers dozens of names for them: preons, rishons, haplons, glicks.... One of the basic unsolved problems in this field is the construction of very small and very light, practically massless, particles out of subparticles which, at the same time, must be at very short distances. This problem is not encountered either in atoms or in nucleons.

As follows from the uncertainty relation, the levels of a system whose size r is very small must be very widely spaced: $\Delta m \gtrsim 1/r$. The excellent agreement of the magnetic moments of the electron and muon with predictions from quantum electrodynamics shows that $r \lesssim (1 \text{ TeV})^{-1}$. Hence, Δm must be $\gtrsim 1$ TeV. Consequently, the muon cannot be considered an excited state of the electron.

There is also another difficulty: If different particles have common constituents, these particles must transform into each other rather rapidly, for example, $\mu^+ \to e^+ + \gamma$, $p \to e^+ + \pi^\circ$, and so on; we know that this is not realized. And if we increase the number of distinct sorts of subparticles, we do not achieve any simplification in comparison with the standard picture.

It is extremely difficult to predict the future of fundamental physics. The progress of physics appears logical and consistent only in restrospect. But if we look not at "postdictions" but at predictions, each subsequent important step is almost always unexpected and quite often not taken seriously: not only by outsiders but also by the very ones who take the step.

If we nevertheless attempt to think about the future, it seems quite plausible that the next step to the further unification of physics will be possible only as a result of the discovery of some new fundamental principle. In

---

†See Author's note, p. 113.

order to grow simpler, physics must become still more nontrivial. Simplicity will not be simple.

Author's Note (Autumn 1983).

When the translation of this book had already been completed, the UA1 and UA2 Collaborations working at the CERN p$\bar{\text{p}}$ collider reported the observation of the first events of the production and decay of W bosons. The first announcement of the discovery of W bosons was made on 20 January 1983 at a CERN seminar. The respective papers were published as preprints, CERN EP/83-13 21 January (UA1 Collaboration) and CERN EP/83-25 15 February (UA2 Collaboration), and appeared in *Physics Letters* (G.Arnison et al. *Phys.Lett.* **122B** (1983) 103; M.Banner et al. *Phys. Lett.* **122B** (1983) 476). The W bosons were detected by their decays into an electron and a neutrino. The W boson mass was found to be approximately 80 GeV, in agreement with theoretical predictions.

In June 1983, the collaboration UA1 reported the observation of the first events of the creation and decay of Z bosons: four events of decay into a e^+e^- pair and one event of decay into a $\mu^+\mu^-$ pair. The mass of the Z boson was found to agree with the theoretical predictions, being of the order of 95 GeV (G.Arnison et al.: Preprint CERN EP/83-73. *Phys.Lett.* **126B** (1983) 398).

In August 1983, the UA2 collaboration reported the observation of eight Z boson events (P.Bagnaia et al.: Preprint CERN EP/83-112). Within the same month, the UA1 collaboration reported 52 $W^\pm$ boson events. The production cross-sections and angular distribution of decay electrons agree with the theory (G.Arnison et al. Preprint CERN EP/83-111).

In June 1983, the maximum luminosity of the CERN p$\bar{\text{p}}$ collider exceeded 10^{29} cm^{-2}s^{-1}.

Appendix 1

On Systems of Physical Units

The system of units in which $\hbar = c = 1$. Electron volt. Comparison of G_N and G_F. Kelvin. Barn. SI. Powers of ten. Ampere. Coulomb. Volt. Farad. Ohm. Weber. Tesla. SI as a standard system. On the advantages and shortcomings of SI. Bibliography.

An adequate choice of units for describing a certain set of phenomena is a powerful tool of science. If the units are adequate, it is easy to carry out a dimensional analysis of phenomena, to estimate a characteristic scale by the order of magnitude, and to reveal the relation to other, seemingly unrelated, phenomena.

At the same time, unusual units are a hindrance for an outsider trying to read and understand special literature. This Appendix was written especially to help such an outsider. It is mostly devoted to the system of units in which $\hbar = c = 1$ and to comparing it with the International System of Units (SI).

The system of units in which $\hbar = c = 1$

The system $\hbar = c = 1$ is widely used in the physics of elementary particles. This system is convenient because the physics of elementary particles deals with quantum relativistic phenomena and, thus, it is natural to use the quantum of action $\hbar$ for the unit of action and the velocity of light c for the unit of velocity. Since both these quantities are universal constants, it is natural to take the next step and assume them to be unities. (In fact, the velocity of light is assumed to be equal to unity when distances in astronomy are measured in light years; one only has to drop the adjective "light".) This turns velocity v, action S, and angular momentum J into dimensionless quantities: $[v]=[S]=[J]=1$.† The dimensions of the spatial coordinates **r** and temporal coordinate t are identical: $[\mathbf{r}]=[t]$. The dimensions of energy E, momentum **p**, and mass m are also identical: $[E]=[\mathbf{p}]=[m]$. Moreover, if we take into account the quantum-mechanical relation between energy E and frequency ω, $E=\hbar\omega$, or between momentum **p** and wavelength λ of a particle, $\mathbf{p}=2\pi\hbar/\lambda$, then it is obvious that

$$[\mathbf{r}^{-1}]=[t^{-1}]=[\mathbf{p}]=[E]=[m].$$

It is also not difficult to show that, with the units $\hbar=c=1$, we have

$$[\mathbf{A}]=[A_0]=[m] \text{ and } [\mathbf{E}]=[\mathbf{H}]=[m^2].$$

Here **A** is the vector potential, A_0 the electric potential, and **E** and **H** the strength of the electric and magnetic fields, respectively. The dimension of the Lagrangian $\mathcal{L}$ is $[m^4]$. All boson fields, like the photon field, have the dimension $[\psi] = [m]$ and the dimension of all fermions is $[\psi] = [m^{3/2}]$. The easiest way to verify this is to look at the corresponding mass terms in the Lagrangian: $m^2\psi^+\psi$ and $m\bar{\psi}\psi$. Hence, all physical quantities with nonzero dimensions in the system $\hbar=c=1$ can be measured in units of energy or mass.

The electric charge e†† in the system $\hbar = c=1$ is a dimensionless quantity: $e^2/\hbar c=\alpha$, where α is the so-called fine structure constant (this term originated in atomic physics, where α defines the scale of the so-called fine splitting of atomic levels); $\alpha^{-1} = 137.03604(11)$†††. The color charge and weak charge, whose squares are denoted by α_s and α_w, respectively, are dimensionless as well. The Fermi constant of the four-fermion weak interaction, G_F, is a dimensional quantity: $[G_F]=[m^{-2}]$. The Newton constant of gravitational interaction, G_N, has the same dimension.

---

†Square brackets designate the dimension of the enclosed physical quantity.

††The unit electric charge e that we use throughout this Appendix is normalized as follows: $e^2/\hbar c = \alpha$. The electron charge given in tables of physical quantities usually corresponds to just this normalization. In all other parts of the book, we use a different normalization of the unit electric charge: $e^2/4\pi\hbar c = \alpha$. It is this last normalization that is widely used in books and papers dealing with quantum electrodynamics and quantum field theory. In the first case, the Coulomb potential between two electrons is e^2/r; in the second, it is $e^2/4\pi r$.

†††Throughout this Appendix the number in parentheses indicates the uncertainty of one standard deviation in the last significant digits of the number: $137.03604(11) = 137.03604 \pm 0.00011$.

APPENDIX 1. ON SYSTEMS OF PHYSICAL UNITS

Electron volt

The unit of energy in SI is one joule:

$$1 \text{ J} = 1 \text{ kg m}^2 \text{ s}^{-2}.$$

In the CGS system, the unit of energy is one erg:

$$1 \text{ erg} = 1 \text{ g cm}^2 \text{s}^{-2} = 10^{-7} \text{ J}.$$

Elementary particle physics uses the electron volt as the unit of energy; its derivatives are: keV (10^3 eV), MeV (10^6 eV), GeV (10^9 eV), and TeV (10^{12} eV). The unit which is employed especially frequently in the special literature and throughout this book is 1 GeV.

One electron volt is the energy transferred to an electron going through a potential difference of one volt. The electron charge is $1.6021892(46) \times 10^{-19}$ C; hence, one coulomb (C) contains $6.241459(93) \times 10^{18}$ electrons.

1 Joule = 1 Coulomb × 1 Volt = $6.241459(93) \times 10^{18}$ eV ≈ 6.24 10^9 GeV
1 GeV = $1.6021892(46) \times 10^{-10}$ J = $1.7826759(52)$ 10^{-24} g c^2,

where c is the velocity of light, $c = 2.99792458(1.2) \times 10^{10}$ cm s^{-1},

$$\hbar c = 1.9732858(51) \times 10^{-14} \text{ GeV cm}$$

and

$$\hbar = 6.582173(17) \times 10^{-25} \text{ GeV s} = 1.0545887(57) \times 10^{-27} \text{ erg s}.$$

In the units $\hbar = c = 1$,

1 GeV ≈ 1.6×10^{-10} J ≈ 1.8×10^{-24} g,
1 GeV^{-1} ≈ 0.7×10^{-24} s ≈ 2×10^{-14} cm.

Comparison of G_N and G_F

It will be very instructive to compare G_N and G_F. In the CGS and SI systems,

$G_N = 6.7 \times 10^{-8}$ cm^3 g^{-1} s^{-2} = 6.7×10^{-11} m^3 kg^{-1} s^{-2},
$G_F = 1.4 \times 10^{-49}$ erg cm^3 = 1.4×10^{-62} J m^3.

A cursory look at these numbers may give the impression that G_N is much greater than G_F. If, however, we convert them to natural units, we immediately realize that the opposite is true:

$G_N \simeq 6.7 \times 10^{-39}$ $\hbar c^5$ GeV^{-2},
$G_F \simeq 1.2 \times 10^{-5}$ $\hbar^3 c^3$ GeV^{-2}.

We thus find that, in $\hbar=c=1$ units, the gravitational constant is weaker than the Fermi constant by 33 orders of magnitude. This is in perfect agreement with the well-known fact that, under laboratory conditions, the gravitational interaction is negligibly small compared to the weak interaction.

Kelvin

Since the absolute temperature T characterizes the mean energy of an ensemble of particles, it is natural to measure temperature in electron volts as well. We then have to write T instead of kT. The Boltzmann constant k is simply the factor for converting from Kelvin degrees (K) to energy units: $k \sim 1$ eV/11604 K. If we take $k=1$, then 1 eV $\simeq$ 11604 K.

The Stefan-Boltzmann constant in the units $\hbar=c=k=1$ becomes

$$\sigma = \pi^2/60.$$

Barn

Cross sections are measured in nuclear physics and the physics of elementary particles in barns (1 b = 10^{-24} cm^2), millibarns (1 mb = 10^{-27} cm^2), microbarns (1 μb = 10^{-30} cm^2), nanobarns (1 nb = 10^{-33} cm^2), picobarns (1 pb = 10^{-36} cm^2), femtobarns (1 fb = 10^{-39} cm^2), and attobarns (1 ab = 10^{-42} cm^2):

$$1 \text{ GeV}^{-2} = 0.389 \text{ mb}.$$

SI

SI is the abbreviation for the French term Système International d'Unitées)—the International System of Units. The basic mechanical units of SI are the meter, m; kilogram, kg; and second, s. The secondary mechanical units of SI are:

force: newton, N = kg m s^{-2};
energy: joule, J = N m;
power: watt, W = J s^{-1},
pressure: pascal, Pa = N m^{-2},
frequency: hertz, Hz = s^{-1}.

APPENDIX 1. ON SYSTEMS OF PHYSICAL UNITS

The basic electromagnetic unit of SI is the ampere (A). Derived electromagnetic units are coulomb (C), volt (V), farad (F), ohm (Ω), weber (Wb), and tesla (T).

Powers of ten

Powers of ten are denoted in SI by means of the following prefixes:

10^{-1}	deci	d	10^{1}	deca	da
10^{-2}	centi	c	10^{2}	hecto	h
10^{-3}	milli	m	10^{3}	kilo	k
10^{-6}	micro	mc, μ	10^{6}	mega	M
10^{-9}	nano	n	10^{9}	giga	G
10^{-12}	pico	p	10^{12}	tera	T
10^{-15}	femto	f	10^{15}	peta	P
10^{-18}	atto	a	10^{18}	exa	E

Ampere

The ampere (A) is defined in SI as the unit of constant electric current that, flowing through two parallel infinitely long conductors placed in vacuum at a distance of 1 m from each other, exerts on them a force equal to 2×10^{-7} newton (N) per each meter of length.

If we write the Ampere's law in the form

$$F = 2I^2 l/c^2 d,$$

where F is the force, I is the current, l is the length of the conductors, d is the distance between them, and c is the velocity of light, then 1 A = $c \sqrt{10^{-7} N}$. But the ampere is treated in SI as a basic, not as a derived unit. Ampere's law is written in SI in the form

$$F = \frac{\mu_0}{4\pi} \frac{I^2 l}{d},$$

where μ_0 is the magnetic permeability of vacuum, expressed through the unit of inductance, the henry (1 H = 1 m² kg s^{-2}A^{-2} = 1 Wb/A; for the definition of the weber (Wb), see below),

$$\mu_0 = 4\pi \times 10^{-7} \text{H/m} = 4\pi \times 10^{-7} \frac{\text{N}}{\text{A}^2}.$$

Coulomb

The coulomb (C) is defined in SI as $1\,A \times 1\,s$. Let us trace the relation of the unit of the amount of electricity, defined in this manner, to another definition based on Coulomb's law. It is convenient to turn to the CGS system of units (cm g s) and define the electrostatic unit of electricity (esu) by the relation

$$(esu)^2/cm = erg.$$

Hence, $(esu)^2 = g\,cm^3 s^{-2} = 10^{-9}\,kg\,m^3\,s^{-2} = 10^{-9}\,N\,m^2$. While comparing it to the definition of the coulomb in SI we find that

$$C^2 = A^2\,s^2 = 10^{-7}\,N\,c^2\,s^2 = 10^{-7}\,N\,m^2(c\,\tfrac{s}{m})^2 = 10^2\,(esu)^2(c\,\tfrac{s}{m})^2\,.$$

Therefore,

$$1C = \tfrac{1}{10}(esu)(c\,\tfrac{s}{m}) = esu\,\tfrac{1}{10}\zeta.$$

where $\zeta = 2.99792458(1.2) \times 10^{10}$ is the numerical value of the velocity of light in vacuum, c, measured in cm s^{-1}. The coefficient 10 is the corollary of the historical coefficient 10^{-7}.

In SI, Coulomb's law is written in the form

$$F = \frac{1}{4\pi\epsilon_0}\frac{Q^2}{r^2},$$

where Q is the electric charge in coulombs, r is the spacing between the charges in meters, and ϵ_0 is the dielectric constant of the vacuum, expressed through the unit of capacitance, the farad (1 F = $1\,m^{-2}kg^{-1}s^4 r_{\!*}^2$ = 1C/V; for the definition of the farad and the volt (V), see below):

$$\epsilon_0 = (4\pi\zeta^2)^{-1} \times 10^{11}\,F/m = 8.85418782(7) \times 10^{-12}\,F/m.$$

Obviously,

$$\epsilon_0\mu_0 = \frac{10^4\,FH}{\zeta^2 m^2} = \frac{10^4}{\zeta^2}\left(\frac{s}{m}\right)^2 = \frac{1}{c^2}.$$

Volt

The volt (V) is the unit of electric voltage, electric potential, or electromotive force (emf). According to SI, 1 volt=1 joule/1 coulomb. Hence,

$$1\text{V} = \frac{\text{erg}}{\text{esu}}\left(\frac{10^8\text{cm/s}}{c}\right) = 10^8\frac{\text{erg}^{1/2}}{\text{cm}^{1/2}}\left(\frac{\text{cm/s}}{c}\right) = \frac{\text{J}}{\text{esu}}\left(\frac{10}{\zeta}\right).$$

As the practical unit of emf, the volt was enacted at the above mentioned Congress of 1881. The Congress defined the volt as 10^8 units of the CGSM system: 1 volt=$10^8(\frac{\text{erg}}{\text{cm}})^{1/2}(\frac{\text{cm/s}}{c})$. Note in passing that the joule was introduced into the system of absolute practical electric units as a unit of work and energy at the Second International Congress of Electricity (1889).

The reader will easily notice that the coefficient 10^8 in the definition of the volt in terms of erg and esu gives a coefficient 10 in the definition of the coulomb in terms of esu and, hence, the coefficient 10^{-7} in the definition of the ampere in terms of the newton. How then was the coefficient 10^8 chosen in the definition of the volt? The answer is that the volt thus defined is close to the emf of the so-called normal cells.

Normal cells are galvanic cells whose emf is stable in time and technologically reproducible. There was a time when normal cells served as primary references of one volt. They are still widely used at the present time to standardize voltage in instrumentation and industry.

Farad

The farad (F) is the unit of electric capacitance. $1\text{ F} = 1\text{C}/1\text{V} = \text{m}^{-2}\text{kg}^{-1}\text{s}^4\text{A}^2$. Recalling that

$$\text{A}\times\text{s} = \text{C} \text{ and } \text{C}^2 = 10^{-7}\text{N } c^2 \text{ s}^2$$

and converting to the system CGS, we readily obtain

$$1\text{F} = 10^{-7}\frac{\text{N } c^2\text{s}^2}{\text{kg m}^2} = 10^{-7}\frac{c^2\text{s}^2}{\text{m}} = 10^{-9}\text{cm}\left(\frac{c\text{s}}{\text{cm}}\right)^2$$

$$= 10^{-9}\zeta^2\text{cm} \simeq 9\times 10^{11}\text{cm}.$$

Ohm

Resistance is measured in ohms (Ω).

$$1 \text{ ohm} = 1\text{V}/1\text{A} = \text{J s C}^{-2} = \text{J s (esu)}^{-2}\left(\frac{10 \text{ cm s}^{-1}}{c}\right)^2$$

$$= 10^9 \frac{\text{s}}{\text{cm}}\left(\frac{\text{cm s}^{-1}}{c}\right)^2 = 10^9 \frac{1}{c}\left(\frac{\text{cm s}^{-1}}{c}\right) = \frac{10^9}{\zeta c}$$

Hence, the ohm has the dimensions of (velocity)$^{-1}$. In a system with $c=1$, 1 ohm$\simeq 1/30$.

Weber

According to SI, the magnetic flux (flux of magnetic induction) is measured in webers: 1 weber=1 joule/1 ampere=1 volt second.

Tesla

The magnetic flux density (magnetic induction) is measured in teslas (T):

1 tesla = 1 weber/m^2 = kg s^{-2}A^{-1} = V s m^{-2},
1 gauss = 10^{-4} tesla,
e tesla m = eVsm^{-1}

where e is the electron charge.

As an example, consider the motion of an electron with momentum p GeV/c along an orbit with radius ρ meters in a transverse magnetic field with strength H teslas. In SI units,

$$p = e\text{H}\rho.$$

Then, by multiplying the left- and right-hand sides by c, we find

$$pc \text{ (GeV)} \simeq 0.3 \text{ H}\rho \text{ (tesla m)}.$$

Here we have taken into account that e×1 tesla m = e×1 V s m^{-1} = 1 eV s m^{-1} and have used the approximate equality $10^{-9}c$ s $m^{-1} \simeq 0.3$.

SI as a standard system

Transition to SI was recommended by the XIth General Conference on Measures and Weights (1960). In the USSR, the decree of the USSR GOSSTANDART of 25 June 1979 introduced the State Standard ST SEV

APPENDIX 1. ON SYSTEMS OF PHYSICAL UNITS   123

1052-78 "Metrology. Units of physical quantities" 1980, to become effective as a basis for contractual and legal relations and in the national economy as of 1 January 1980.

On the advantages and shortcomings of SI

The main advantage of SI is that most SI units are convenient for practical applications in many fields of science and technology. The diversity of traditionally formed units for each physical quantity is replaced by a single unit and a clearcut system of forming multiple and fractional units out of that one is introduced.

The main shortcoming of SI is that, in fundamental physics, many formulas written in SI take on cumbersome conversion factors of not so much physical as historical nature, which hampers understanding of the essence of physical phenomena. This stems, first of all, from the facts that the strengths of the electric and magnetic fields in vacuum, **E** and **H**, obtain different dimensions in SI and that the vacuum is assigned nonunity dimentional magnetic permeability μ_0 and dielectric constant ϵ_0; as a result, the dimensions of all four vectors, two strengths, **H** and **E**, and two inductions, $\mathbf{B}=\mu_0\mathbf{H}$ and $\mathbf{D}=\epsilon_0\mathbf{E}$, are different. Such definitions correspond to the prerelativistic level of electrodynamics of about a century ago and sharply contradict the physical meaning of Maxwell's equations, for instance, that **E** and **H** are the components of one and the same four-dimensional tensor of the electromagnetic field, $F_{\mu\nu}$. A natural desire to retain practical units: the coulomb, volt, and joule, is no justification for these shortcomings of SI. Indeed, it is possible to combine both these units and meaningful definitions of the vacuum and the strength of the electromagnetic fields.

In general, an all-encompassing mandatory standardization based on a single system of physical units, even if it is more consistent than SI, can only be harmful. We have shown above that understanding is much facilitated in elementary particle physics if we use the system of units $\hbar=c=1$. But one would be stupid to insist that this system be employed in, say, daily life or agriculture.

The State Standard cited above recognizes that a reasonable choice of physical units is required for scientific research. A special reservation was put into this standard that it "does not cover the units employed in scientific research and publications of a theoretical nature."

But this is only a half-hearted recognition because, at the same time, the text of the standard declares that the "educational process (including textbooks and other aids) must be based in all educational establishments on the use of SI units and units admitted for use as stipulated by subsections 3.1, 3.2, 3.3." Note that neither the CGS system nor the $\hbar=c=1$ system widely used in physics are even mentioned in these subsections. But is it admissible to cut the link between science and education? A student who misses the physical meaning of Maxwell's equations today will be unable to work with them tomorrow.

It is commendable that the standard permits the use of the electron volt in physics along with the joule. But dropping the unit of cross section, the barn (10^{-24} cm^2), and its fractional units is extremely unfortunate, all the more so because SI has no prefixes denoting fractions smaller than 10^{-18}.

Several years ago, hectopascals were introduced resolutely and suddenly into weather forecasts on the radio and in the papers, replacing the nonstandardized millimeters of mercury column. However, this innovation had to be abandoned several weeks later. It was obvious to everyone how unprepared, untimely, and harmful the change was. The harm done to science and, in the long run, to technology, by an immoderate and inadequate standardization is obvious to a narrow circle of specialists. But this harm is incomparably more serious.

Bibliography

1. Symbols, Units and Nomenclature in Physics: Document U.I.P. 20 (1978). International Union of Pure and Applied Physics, S.U.N. Commission.
2. Sivukhin D.V.: On the international system of physical units. *Soviet Physics Uspekhi* **129**, no. 2 (1979) 335-338.
3. Leontovich M.A.: On systems of measures. *Vestnik AN SSSR*, no. 6 (1964) 123-126. (Leontovich's and Sivukhin's papers present a criticism of SI. The paper by Sivukhin presents the stand taken by the Division of General Physics and Astronomy of the USSR Academy of Sciences.)
4. Kamke D. and Krämer K.: Physikalische Grundlagen der Masseinheiten. Stuttgart, Teubner, (1977).
5. Cohen E., in: Enrico Fermi Summer School, course entitled "Metrology and Fundamental Constants". Amsterdam, North-Holland, (1976).
6. Cohen E.R. and Taylor B.N.: *J. Phys. Chem. Ref. Data* **2**, no. 4 (1973) 663.

7. Hooper W. *Am. J. Phys.* **48** (1980) 681. (A survey of the use of the International System of Units (SI) in college physics texts in the USA.)
8. Hooper W *Am. J. Phys.* **51** (1983) 683; Bueche F. *Am. J. Phys.* **51** (1983) 684. (Letters to the editor discussing the use, pseudouse, and misuse of SI units instead of such units as BTU, mm Hg, kilocalories, etc. The natural units $\hbar = c = 1$ are not mentioned in this exchange of letters.)

Appendix 2

GLOSSARY

This glossary contains about 120 terms. Most of them come from elementary particle theory and mathematics, but not exclusively; thus, a number of entries are devoted to accelerators. The idea was to assign several parallel functions to the glossary:

1. to supplement and clarify the text of the Survey,
2. to serve as a reference source irrespective of the Survey,
3. to help the beginner to find his way through the terminology. The terminological barrier can be one of the toughest obstacles blocking off a new field. Simple and brief definitions of terms, with etymological clues, will help, at least partially, to overcome this barrier.
4. to warn against using obsolete terms such as, for example, "rest mass" and "mu meson."
5. to give examples of simple dimensional estimates based on the system of units $\hbar = c = 1$.

Obviously, a glossary with such diverse functions could not be uniform in form and style. I doubt that cropping it to a universal crew cut would do much good. I did try to write the opening sentences of each entry in a simple and clear manner. Sometimes the entry as a whole is also simple, but a number of passages are meant for a professional theorist. If you feel difficulty in understanding, you are probably going through just such a passage. Skip it.

Quite a few important terms will not be found among the headings of glossary entries. In most cases, they are explained in other related entries or in the body of the Survey. They can be quickly located if you address the Subject Index.

Annihilation. A process by which a colliding particle and antiparticle annihilate one another, thereby creating other particles. Historically, the first observed annihilation process was the annihilation of an electron and a positron into two photons: $e^+e^- \to 2\gamma$. The most interesting at present is the annihilation of an electron and a positron into hadrons at high energies in colliding electron-positron beams.

Anomaly. In quantum field theory, the violation of a conservation law in the case where the law results from some symmetry of the Lagrangian and is violated by quantum corrections calculated on the basis of the same Lagrangian. This paradox occurs because some Feynman diagrams are ill-defined at infinitely high momenta of virtual particles and therefore call for a special regularization procedure. It is this procedure that contradicts the original symmetry of the Lagrangian.

The best known example of an anomaly is the nonconvervation of the axial current of a massless charged fermion. Formally, the Lagrangian is chirally invariant in this case; nevertheless, there is a so-called triangular diagram (see Figure 42) that violates the conservation of axial current. The sides of the triangle in Figure 42 represent the propagation of a virtual fermion ("electron"). The upper vertex represents the coupling of the fermion axial current to a Z boson and the two lower vertices stand for the electromagnetic coupling of the fermion vector current to a photon.

Although the contribution of a triangular diagram to the divergence of the axial current is finite, its contribution to the axial current as such is ultraviolet-divergent and has to be regularized at high momenta of virtual particles. It has been found to be impossible to carry out a meaningful regularization that would not violate the conservation of the axial current.

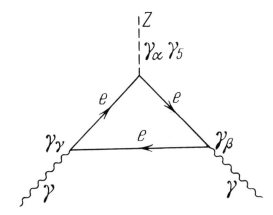

FIGURE 42. The axial anomaly: the triangular diagram which violates the conservation of the axial current.

APPENDIX 2. GLOSSARY

The requirement that the theory must be free of axial anomalies imposes a serious restriction on the admissible structure of fermion multiplets. Let us mention, as an example, that the SU(5) model of grand unification is anomaly-free because the quark and lepton anomalies exactly cancel out.

Another important example is the anomaly in the trace of the energy-momentum tensor. This anomaly plays an important part in quantum chromodynamics.

In gluodynamics, the Lagrangian of the massless gluon field,

$$\mathcal{L} = \frac{1}{4}F^a_{\mu\nu}F^a_{\mu\nu}, \qquad a = 1, 2, \ldots, 8$$

is scale invariant. We could thus naively expect that, in this case, the trace of the energy-momentum tensor $\theta_{\mu\mu}$ would be zero. (Roughly speaking, $\theta_{\mu\nu} \sim p_\mu p_\nu$, where p is the 4-momentum of a gluon, and, since $p^2 = 0$ for a massless gluon, $\theta_{\mu\mu} = 0$.) However, when the triangular diagram in Figure 43 is regularized, we find that $\theta_{\mu\mu} \neq 0$. (The upper vertex in the diagram of Figure 43 represents the coupling of the gluon to the graviton and the two lower vertices stand for the nonlinear gluon-gluon coupling.) Calculations show that

$$\theta_{\mu\mu} = \frac{1}{2}\frac{\beta(g_s^2)}{g_s^2}F^a_{\mu\nu}F^a_{\mu\nu},$$

where $\beta(g_s^2)$ is the so-called Gell-Mann–Low function and $g_s^2 = 4\pi\alpha_s$ is the dimensionless gluon coupling constant; $\beta(g_s^2) = -(4\pi)^{-2}bg_s^4$ + terms of higher order in α_s. In the case of gluodynamics (SU(3) symmetry), $b = 11$.

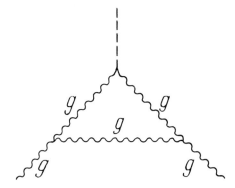

FIGURE 43. The gluonic triangular diagram responsible for the anomaly in the trace of the energy-momentum tensor in quantum gluodynamics. Wavy lines represent gluons and the dashed line represents the graviton.

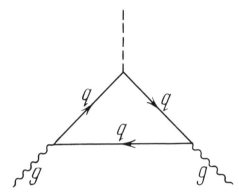

FIGURE 44. The triangular anomaly due to massless quarks in quantum chromodynamics. Solid lines: quarks; wavy lines: gluons; dashed line: the graviton.

The trace of the energy-momentum tensor equals the divergence of the so-called dilatation current K_μ ($K_\mu = \nu\theta_{\nu\mu}$). If $\theta_{\mu\mu} \neq 0$, the scale invariance is broken and the dilatation current is not conserved.

The appearance of a scale (dimensional constant Λ_{QCD}) in gluodynamics is thus closely related to the anomaly in the trace of the energy-momentum tensor. Asymptotic freedom and confinement are also closely associated with this anomaly. Massless fermions also contribute to the anomaly of the energy-momentum tensor (see Figure 43, where the triangle represents the contribution of a massless quark). In quantum chromodynamics, where not only gluons, but also light quarks have to be taken into account, we find:

$$b = 11 - \frac{2}{3}n_f,$$

where f is the number of flavors of light quarks.

Antiparticle (with respect to a given particle). A particle with mass, spin, and lifetime identical to those of the given particle but with opposite signs of all charge-like quantum numbers. The term "charge-like quantum number" here means electric charge, lepton quantum number (sometimes called lepton charge), baryon quantum number (sometimes called baryon charge), hypercharge, color charge, and so forth. For instance, the positron is the antiparticle of the electron and the electron is the antiparticle of the positron. A similar pair is formed by the proton and antiproton. The neutron and antineutron are both electrically neutral but differ in the signs of the baryon number and of the magnetic moment. The antiparticle of the Σ^+ hyperon is not the Σ^- hyperon, which has the same sign of the baryon number, but the $\tilde{\Sigma}^-$ hyperon (sometimes denoted by $\tilde{\Sigma}^+$).

The antiparticle is usually designated by the same letter as the particle with a tilde or bar over it. This notation is not completely satisfactory because it does not conform with the completely symmetric roles of a particle and its antiparticle and with the relative sense of the very notion "antiparticle." However, a better notation has not been devised.

Particles with no charge quantum numbers are said to be truly neutral. They are their own antiparticles. Examples of truly neutral particles are the photon, π° meson, η meson, η' meson, ϕ meson, levels of charmonium, and levels of upsilonium.

Asymptotic freedom. In QCD, a logarithmic weakening of the strong (color) interaction between quarks and gluons with decreasing distance between them. Asymptotic freedom is inherent in the gauge interaction in quantum chromodynamics and is generally inherent in non-Abelian gauge theories; experimentally it manifests itself in deep inelastic processes and in the properties of heavy quarkonia. In mathematics, "asymptotic" means approaching infinitely closely (from the Greek $\alpha\sigma\upsilon\mu\pi\tau\omega\tau\sigma\varsigma$, "not meeting").

The property of the asymptotic freedom of non-Abelian gauge theories was discovered in 1973 by D. Gross, F. Wilczek, and H.D. Politzer. This property originates in the logarithmic dependence of the coupling constant g (the so-called invariant charge) on the square of momentum transfer q^2 (here and below, $q^2 = \mathbf{q}^2 - q_0^2$):

$$\frac{dg^2(q^2)}{d\ln q^2} = \beta(g^2),$$

where $\beta(g^2)$ is the Gell-Mann–Low function. In quantum chromodynamics,

$$\beta(g_s^2) = -b \frac{g_s^4}{16\pi^2} + \text{terms of higher orders in } g_s^2/4\pi.$$

Here $b = 11 - \frac{2}{3} n_f(q^2)$ is the number of quark flavors satisfying the condition $4m^2 \ll q^2$. In standard chromodynamics with six quark flavors, this relation yields the asymptotic

$$\alpha_s(q^2) \equiv \frac{g_s^2(q^2)}{4\pi} \sim \frac{4\pi}{7\ln q^2/\Lambda_{QCD}^2}$$

as q^2 approaches infinity. Here Λ_{QCD} is, by definition, that value of momentum at which α_s becomes infinite (obviously, the above asymptotic formula for α_s is not valid for such small q). The expression for b shows that asymptotic freedom is broken if the number of quark flavors is sufficiently large ($n_f > 16$).

For an arbitrary gauge group G with Dirac fermions in the representation R_f and with complex scalars in the representation R_s,

$$b = \frac{11}{3}C(G) + \frac{4}{3}T(R_f) + \frac{1}{3}T(R_s).$$

Here

$$C(G)\delta^{ab} = t^{acd}t^{bcd},$$
$$T(R)\delta^{ab} = tr\{\Gamma^a, \Gamma^b\},$$

where $C(G)$ is the Casimir quadratic operator for the adjoint representation of the group G, t^{acd} are the structure constants of the group, and Γ^a are the generators of the representation R. For instance, $C = N$ for the adjoint representation of the group $SU(N)$ and $T = \frac{1}{2}$ for the spinor representation of the same group.

Axial current (or axial vector current). A current with the transformation properties of a four-dimensional axial vector. The total weak current is the sum of a vector and an axial current. The product of a vector and an axial current is a pseudoscalar; it is the source of parity violation in weak interactions.

Axial vector. The same as a pseudovector. In contrast to the components of the ordinary (polar) vector, the components of an axial vector do not change sign under mirror reflection (inversion) of spatial coordinates. The vector product of two polar vectors is an axial vector. Examples of physical quantities represented by axial vectors are the magnetic field strength **H**, angular momentum **J**, and axial and the spatial part of weak current.

A three-dimensional axial vector **H** is described by three spatial components of a six-component four-dimensional asymmetric tensor $\mathbf{F}_{\mu\nu}$.

In the case of axial currents, a three-dimensional axial vector is described by the three spatial components of a four-dimensional axial vector whose time component behaves as a pseudoscalar under spatial rotations and reflections.

Axion. A hypothetical neutral spinless light particle (with mass less than 1 MeV) that interacts very weakly with matter and decays into two photons. The axion hypothesis was proposed in relation to the problem of the conservation of CP parity in strong interactions (the so-called problem of the θ term) and the related problem of the conservation of the Abelian axial current (hence the term "axion"). The existence of a physically observable axion is not an obligatory corollary of the theory.

Bag model. A phenomenological model of hadrons that treats them as peculiar bubbles, or "bags," in which confined quarks move freely. The bag model, operating with several adjustment parameters, gives a fairly good spectrum of hadron masses. From the viewpoint of quantum chromodynamics, a bubble in the chromodynamical vacuum is formed because color fields of quarks destroy, or rather "burn out" (totally or partially), the gluon condensate in the physical vacuum. Note that the energy of the bubble is positive because the gluon condensate in the physical vacuum has negative energy: A hole in the negative background is positive.

Baryons. Hadrons with half-integral spin. All baryons have a baryon quantum number (baryon charge) equal to $+1$; for all antibaryons, it is equal to -1. The baryon quantum number is conserved in strong, weak, and electromagnetic interactions. Consequently, all processes due to these interactions conserve the difference between the numbers of baryons and of antibaryons. Grand unification models predict processes in which baryon charge is not conserved (proton decay). The word baryon comes from the Greek word $\beta\alpha\rho\upsilon\varsigma$, "heavy").

Baryonium. A system consisting of a baryon and an antibaryon coupled by strong attractive force.

b hadrons. Hadrons comprising b quarks. Sometimes b hadrons are referred to as beautiful particles. The lightest of b hadrons are $B^- = b\bar{u}$, $B^+ = \bar{b}u$, $\bar{B}^0 = b\bar{d}$, and $B^0 = \bar{b}d$; their masses are approximately equal to 5.274 GeV, within an uncertainty of a few MeV. The lightest of b baryons, $\Lambda_b = udb$, is somewhat heavier (5.5 GeV). Mesons with hidden "beauty," consisting of a b quark and a b antiquark ($\bar{b}$), form the levels of upsilonium. Upsilonium is often called bottonium or bottomonium. The terminology has not yet settled in this field. Sometimes b hadrons are understood only as particles with "naked beauty."

Bjorken scaling. characterizes deep inelastic collisions of leptons with hadrons; it states that deep inelastic formfactors (certain dimensionless functions characterizing the differential cross-sections of these processes) are functions of a single dimensionless variable $x = -q^2/2\nu$. Here q^2 is the square of the four-dimensional momentum q transferred from a lepton to a nucleon and $\nu=qp$, where p is the 4-momentum of the nucleon. In the laboratory frame of reference in which the nucleon is at rest, $p = m$, where m is the nucleon mass, and $\nu/m = E - E'$, where E is the energy of the initial and E' that of the final lepton. Hence, in the laboratory frame of reference, ν/m is the energy transferred from the lepton to the nucleon. Both the Bjorken scaling as such and the experimentally observed departures from it are explained by quantum chromodynamics.

Blackbody radiation. Photons emitted by a so-called blackbody. A blackbody is defined as a body that absorbs all photons incident on it. In its turn, the blackbody emits photons, the intensity and characteristic frequency of the radiation being determined by the temperature T of the blackbody.

The energy density of a photon gas in thermal equilibrium with a blackbody is $\rho = 4\sigma T^4$, where $\sigma = \pi^2/60$ is the Stefan-Boltzmann constant. Here we use the system of units $\hbar = c = 1$ and set the Boltzmann constant k to unity as well. In these units, the dimension of energy (and frequency) is T and that of volume is T^{-3}, so that the above dependence T^4 is obtained directly from dimensional arguments. The energy flux emitted by a unit area of a blackbody per unit time is $j = \sigma T^4$.

The spectral density of blackbody radiation, $d\rho(\omega)$, is given by the Planck radiation formula,

$$d\rho(\omega) = \frac{\omega^3 d\omega}{\pi^2(e^{\omega/T} - 1)} = \frac{T^4}{\pi^2} \frac{x^3 dx}{e^x - 1}.$$

Here ω is the radiation frequency and $x = \omega/T$. Integration of this expression with respect to x from 0 to ∞ gives

$$\rho = \frac{\pi^2}{15} T^4 = 4\sigma T^4.$$

Black hole. A cosmic object whose gravitational field is produced by a mass m confined to a region whose size is less than the gravitational radius r_g. For a mass m, the gravitational radius r_g is given by the formula $r_g = 2G_N m/c^2$, where G_N is the Newton gravitational constant. In the system of units $\hbar = c = 1$, we have

$$r_g = 2G_N m = 2m/m_P^2,$$

where m_P is the Planck mass. (It is readily verifiable that, for instance, for our Sun, whose mass is 2×10^{33}g, the gravitational radius $r_g \approx 3$ km.)

Neither photons nor other particles trapped by a black hole can overcome classically its strong gravitational attraction and have to stay within the sphere with the radius r_g. Nevertheless, they are emitted off the surface of black holes due to quantum effects. Hawking has shown that a sphere with radius r_g radiates like a blackbody with temperature $T = 1/(4\pi r_g)$. If we take into account only emitted photons, the black hole loses mass at a rate $dm/dt \approx -\sigma T^4 4\pi r_g^2$, where $\sigma = \pi^2/60$ is the Stefan-Boltzmann constant. This gives

$$\frac{dm}{dt} \simeq -\frac{m_P^4}{15 \times 2^{10}\pi m^2}.$$

The lifetime of a black hole is then

$$t = 5 \times 2^{10}\pi m^3 m_P^{-4},$$

so the lifetime of black holes with $m > 10^{15}$g exceeds that of the universe. The coefficients in the above expression should not be regarded as exact because a number of effects were ignored. Numerical calculations for black holes with $m \gg 10^{17}$g (Page, 1976) give $dm/dt \cong -2 \times 10^{-4} m_P^4 m^{-2}$. Of this, 81 percent is carried away by ν_e and ν_μ (ν_τ were neglected in the calculation), 17 percent by photons, and 2 percent by gravitons.

Bosons. Elementary or composite particles with integral spin. Bosons are governed by the Bose-Einstein statistics. An arbitrarily large number of bosons of a given type can simultaneously occupy a given quantum state: Examples of bosons are the photon, mesons, the nucleus of ^{4}He, and an atom of this isotope of helium. The word boson derives from the name of the Indian physicist J.C. Bose (1894–1974).

CERN (Organisation Européenne pour la Recherche Nucléaire, formerly Conceil Européenne). European Organization for Nuclear Reserach, near Geneva, Switzerland.

Charge conjugation. An operation replacing particles by antiparticles. Charge conjugation transforms truly neutral particles into themselves.

Charge conjugation parity (C parity). A quantum number characterizing truly neutral particles or systems of particles. If a charge conjugation operation does not reverse the sign of the wave function of a particle, its charge parity is positive ($C = +1$); if it does, the charge parity is negative ($C = -1$). The charge parity of positronium levels (e^+e^-) is a function of orbital momentum L and total spin S of the electron and positron: $C = (-1)^{L+S}$. The same is true for the levels of quarkonium (a quark-antiquark system). The charge parity of the photon and the ρ°, ω, ϕ, J/ψ, and Υ mesons is negative. The charge parity of the π°, η, and η' mesons is positive. Charge parity is conserved in strong and electromagnetic interactions, and its conservation is violated in weak interactions.

Charged currents. Weak currents of leptons and quarks coupled to charged intermediate bosons $W^\pm$. The interactions between (couplings of) charged currents realized by $W^\pm$ bosons cause all known weak decays and a number of weak reactions involving the neutrinos. All known lepton and quark charged currents have the form $\bar{a}O_\alpha b$. Here $\bar{a}$ is the creation operator of a particle a (and annihilation operator of its antiparticle $\tilde{a}$), b is the annihilation operator of a particle b (and creation operator of its antiparticle $\tilde{b}$, and $O_\alpha = \gamma_\alpha(1 + \gamma_5)$), where γ_α are four Dirac matrices ($\alpha = 0, 1, 2, 3$) and $\gamma_5 = i\gamma_0\gamma_1\gamma_2\gamma_3$. The quantity $\bar{a}\gamma_\alpha b$ transforms as a Lorentz vector and $\bar{a}\gamma_\alpha\gamma_5 b$ as a Lorentz axial vector. The total charged current is the sum of lepton and quark currents. The lepton charged current is the sum of three terms: $\bar{\nu}_e O_\alpha e + \bar{\nu}_\mu O_\alpha \mu + \bar{\nu}_\tau O_\alpha \tau$. Not all terms of the quark charged current are yet known. Assuming the existence of six quarks, the quark current is written as the sum of three terms:

$$\bar{u}O_\alpha d' + \bar{c}O_\alpha s' + \bar{t}O_\alpha b',$$

where summation over color is meant in each term, for instance,

$$\bar{u}O_\alpha d' = \bar{u}_1 O_\alpha d'_1 + \bar{u}_2 O_\alpha d'_2 + \bar{u}_3 O_\alpha d'_3,$$

where 1, 2, and 3 are three color indices. The "rotated" quarks d', s', and b' are linear combinations of the quarks d, s, and b, defined by the so-called Kobayashi-Maskawa matrix,

$$\begin{pmatrix} d' \\ s' \\ b' \end{pmatrix} = \begin{pmatrix} c & s_1 c_3 & s_1 s_3 \\ -s_1 c_2 & c_1 c_2 c_3 - e^{i\delta} s_2 s_3 & c_1 c_2 s_3 + e^{i\delta} s_2 c_3 \\ s_1 s_2 & c_1 s_2 c_3 - e^{i\delta} c_2 s_3 & -c_1 s_2 s_3 + e^{i\delta} c_2 c_3 \end{pmatrix} \begin{pmatrix} d \\ s \\ b \end{pmatrix}.$$

Here we use the contracted notations $s_1 = \sin\theta_1$, $c_1 = \cos\theta_1$, $s_2 = \sin\theta_2$, $c_2 = \cos\theta_2$, $s_3 = \sin\theta_3$, and $c_3 = \cos\theta_3$. Among the four parameters θ_1, θ_2, θ_3, and δ, so far only the angle θ_1 is known very well.

Charmonium. The set of mesons consisting of a charmed quark and charmed antiquark ($c\bar{c}$). All these mesons (the so-called mesons with hidden charm) are various energy levels of charmonium. Like positronium levels, charmonium levels are characterized by the following quantum numbers: J, L, S, P, C, and n_r, where J is the total angular momentum, L is the orbital momentum of the quark and antiquark, S is their total spin, P is the spatial parity of level, C is its charge parity, and n_r is the radial quantum number. As in the case of ordinary atoms and positronium, levels with $L = 0, 1, 2, 3, \ldots$ are referred to as S, P, D, F, $\ldots$ levels. Mesons with $S = 1$ are said to be the levels of orthocharmonium and those with $S = 0$ the levels of paracharmonium. The P-parity of a level is given by the relation $P = (-1)^{L+1}$ and the C parity by $C = (-1)^{L+S}$. The lowest state of orthocharmonium with $L = 0$ ($^3S_1$) is called the J/ψ meson. The radial excitations of this state are denoted by ψ', ψ'', ψ''', $\ldots$, where the number of primes equals the number of the radial excitation. The lowest state of paracharmonium with $L = 0$ ($^1S_0$) is called the η_c meson. Orthostates with $L = 1$ and $J = 0, 1, 2$ are denoted by χ_0, χ_1, and χ_2, respectively.

Chiral invariance. Invariance under independent isotopic (or similar continuous-group) transformations of left-handed and right-handed spinors. Quantum chromodynamics would be chiral invariant (its symmetry would be $SU(2)_L \times SU(2)_R$) if the masses of the u and d quarks were zero. In this case, the left-handed and right-handed helicity states of quarks could be subjected to independent isotopic rotations. The sum of generators acting on the left- and right-handed spinors gives the generators of ordinary isotopic rotations. Their difference gives the pseudoscalar generators of isotopic rotations, changing the parity of the states to which these generators are applied.

Since free quarks do not exist and the nucleons containing them are massive, the strict chiral invariance of quantum chromodynamics could be realized only in a nonlinear form. It can be shown that such a nonlinear realization would require massless π mesons. The states with different numbers of zero-momentum π mesons have the same energy, and the above-mentioned pseudoscalar generators transform the states with even and odd numbers of such soft π mesons into one another.

Nature manifests only an approximate chiral invariance because, although the masses of the u and d quarks are small, they are not equal to zero. As a result, π mesons are not massless but are much lighter than other hadrons.

The word chiral derives from the Greek χείρ, "hand."

Classification. The distribution of objects or phenomena among classes. The present book gives numerous examples of different types of classification.

An interesting example of a classification lying far beyond the realm of physics is found in the book *Let Mots et les Choses* by the modern French philosopher Michel Foucault (Gallimard, 1966). Foucault says that he quotes the Argentinean writer Jorge Luis Borges who, in turn, is supposed to quote some Oriental encyclopedia. I was unable to find this quotation in Borges' books, and so give it in the form found in the French text:

'Animals are classified into: (a) those belonging to the Emperor; (b) embalmed; (c) tamed; (d) suckling pigs; (e) sirens; (f) those described in fairy tales; (g) stray dogs; (h) those subsumed into this classification; (i) those who rave as if mad; (j) uncountable; (k) those painted by a very fine camel-hair brush; (l) others; (m) those who have just broken a jug; (n) those who appear from afar as flies.'

If the classification of some physical topic reminds you of this classification of animals, you apparently have not mastered the topic sufficiently well. Study it until the resemblance fades away.

Collaboration. A group of physicists from a number of institutions who jointly conduct an experiment. The largest collaborations run into more than a hundred physicists, and quite a few comprise several dozens of them. Let us mention here the collaborations studying neutrino interactions: CDHS (CERN-Dortmund-Heidelberg-Saclay) and CHARM (CERN-Hamburg-Amsterdam-Rome-Moscow). Sometimes the name of the collaboration is the code name of the experiment, for example, UA1 (Underground Area no.1), the largest experiment run on the CERN proton-antiproton collider.

Confinement. The restraint or incarceration of color quarks and gluons within colorless hadrons. The hypothesis of strict confinement was suggested to explain the negative results of the experimental search for free quarks. Later, as the structure of quantum chromodynamics became clearer, the hope was strengthened that confinement is a corollary of non-Abelian gauge symmetry. This hope is largely based on the results obtained in the framework of the so-called quantum-chromodynamical sum rules, as well as on computer experiments. In these experiments, equations of gluodynamics with infinitely heavy quarks were solved for space-time lattices with a number of cells of the order of 10^4. The results of these nonperturbative computations indicate that the potential between quarks grows linearly with the distance separating them, in accordance with expectations for the gluon string. However, we do not yet have a proof of confinement, nor a clear understanding of this phenomenon in the framework of quantum chromodynamics.

Conformal symmetry. Symmetry under conformal transformations, that is, transformations that leave the angles between directions invariant. In physics, the group of conformal transformations is usually understood as the 15-parameter conformal group of the Minkowski space. The following differential operators represent the generators of this group in the space of scalar functions:

$$\begin{aligned}
M_{\mu\nu} &= x_\mu \partial_\nu - x_\nu \partial_\mu, \\
P_\mu &= \partial_\mu, \\
K_\mu &= 2x_\mu x_\nu \partial_\nu - x^2 \partial_\mu, \\
D &= x_\nu \partial_\nu \qquad (\mu, \nu = 0, 1, 2, 3),
\end{aligned}$$

Here $M_{\mu\nu}$ and P_μ are ten generators of the Poincaré group, which is a subgroup of the conformal group, K_μ are four generators of the so-called special nonlinear conformal transformations, and D is the dilatation generator, which changes the scale.

The necessary condition for the conformal invariance of equations is their scale invariance. The necessary condition for scale invariance of equations is that the Lagrangian contain no dimensional parameters like the masses of particles or the dimensional constants G_F and G_N. For example, the Lagrangian of noninteracting photons and the Lagrangian of noninteracting massless neutrinos are conformally invariant. The chromodynamic Lagrangian describing massless quarks and gluons is also conformally invariant. However, in this last case, quantum corrections violate the conformal invariance (the conformal anomaly).

Cosmic rays. High-energy particles, mostly protons, impinging on the earth from cosmic space (primary cosmic rays) as well as particles created in collisions of primary cosmic rays with atoms in the atmosphere (secondary cosmic rays). Primary cosmic rays are isotropic and time-independent, except for the solar component (the maximum energy of particles in this last component is about 10 GeV). The flux of primary cosmic rays in the range $10-10^6$ GeV for particles with energies above E GeV is approximately equal to $1.7E^{-1.6}$ particles cm^{-2} s^{-1} sr^{-1}. For $E > 10^6$ GeV, the exponent increases from 1.6 to 2.2, so that, for example, particles with energies $E > 10^7$ GeV strike an area of 1 m^2 about once a year.

The positron, muon, π and K mesons, and Λ hyperon were discovered in cosmic rays; the multiple production of hadrons was also found in cosmic rays and the basic features of this phenomenon were established, although the accuracy was not too good. Among other things, some manifestations of the so-called Feynman scaling were discovered.

A number of strange, not yet explained phenomena have been found in cosmic rays, for instance, several so-called Centauro events. These are observed as acts of multiple production at an energy of 10^5–10^6 GeV, in which about one hundred charged particles are created but π° mesons are practically absent. A hunt for Centauros in the CERN proton-antiproton collider was unsuccessful, but this may have been because the energy in the center-of-mass frame was somewhat lower in the experiment than in the cosmic-ray events.

CPT theorem. A fundamental theorem in quantum field theory that states that the equations of the theory are invariant under the product of three transformations: charge conjugation C, spatial inversion P, and time reversal T. If some process takes place in nature, a CPT-conjugate process—in which all particles are replaced with antiparticles, the signs of the spin projections are reversed, and the initial and final states of the process change places—is equally possible. The CPT theorem implies that a particle and its antiparticle have identical masses and lifetimes, their magnetic moments have opposite signs, and they interact identically with the gravitational field. The discoveries of the P- and C-violations (1956), as well as the discovery of the CP-violation (1964), left the theoretical apparatus of physics almost unaltered: It was able to incorporate them in the most natural manner. In contrast to this, if a violation of CPT invariance is ever established experimentally, it would necessitate a radical revision of such fundamental elements of quantum field theory as the causality principle (the locality of the Lagrangian) and the spin-statistics relation.

The CPT theorem was formulated and proved by G. Lüders (1954) and W. Pauli (1955).

Cross section. A quantity with the dimension of area, characterizing the probability of interaction between two colliding particles; usually denoted by σ. The number of interactions N in a beam containing n_1 particles incident on a target plate perpendicular to the beam having area S and n_2/S particles per unit area is given by the formula $N = n_1 n_2 \sigma/S$. Cross section is usually measured in cm^2 or barns (1 barn = 10^{-24} cm^2).

The cross section of the nucleon-nucleon interaction at beam energy in the range 10–100 GeV is approximately 40 mbarn; this corresponds to a nucleon diameter of about 10^{-13} cm, that is, of the order of the confinement radius. The cross section of collisions of π mesons with nucleons in the same energy range is approximately 25 mbarn. These figures give the so-called total cross sections, σ_{tot}, which cover all possible outcomes of a collision, $\sigma_{tot} = \sigma_{el} + \sigma_{inel}$, where σ_{el} is the cross section of elastic scattering and σ_{inel} is the cross section of all inelastic processes, including those in which additional particles are created.

If the direction in which scattered particles emerge is fixed, the cross section is called the differential cross section. The differential cross section of elastic scattering can be given in several different forms. For instance, as $d\sigma/d\Omega$, where $d\Omega = d\varphi d\cos\theta$ is an element of solid angle; or as $d\sigma/d\cos\theta$, if we integrate over φ; or as $d\sigma/dt$, where t is the square of the four-dimensional momentum transfer. In agreement with Gribov's prediction made at the beginning of the 1960s, the values of t characteristic of $d\sigma/dt$ logarithmically decrease as the energy of colliding hadrons increases. This effect is usually called the logarithmic shrinking of the differential cone.

If the momenta of all particles created in a certain channel of an inelastic interaction are identified, then the relevant cross section is called exclusive. If the momentum of only one of the secondary particles is fixed in an inelastic interaction, then the cross section is called inclusive.

Total cross sections of strong interactions between hadrons slowly increase with energy, beginning with an energy of several tens of GeV.

The theory sets a limit on the rate of growth of the total cross section of the strong interaction at asymptotically high energies, the so-called Froissart limit: The cross section cannot increase faster than the squared logarithm of energy. The experimentally observed rise of the cross sections is close to this dependence (although the absolute values of the cross sections are much smaller than the Froissart limit).

The total cross sections of weak four-fermion interactions increase proportionally to E^2, where E is the center-of-mass energy of colliding particles. This behavior can be readily understood in terms of dimensional arguments if we take the dimension of the Fermi constant G_F into account. Indeed, in the $\hbar = c = 1$ system of units, $\sigma \propto G_F^2 E^2$ because $[G_F] = [m^{-2}]$. Note that the growth of the weak interaction cross sections must cease at energies E comparable to the masses of intermediate bosons.

The cross section of electromagnetic annihilation, $e^+e^- \rightarrow$ hadrons, is also easily evaluated on a dimensional basis, $\sigma \sim \alpha^2 E^{-2}$, where $\alpha = 1/137$. Hard weak processes must behave similarly at energies much greater than the masses of intermediate bosons.

Current. A physical quantity determining the interaction between particles and a given vector field. For example, the electromagnetic current determines the interaction between particles and the electromagnetic field. The term of the Lagrangian describing this interaction is $eA_\alpha j^\alpha$, where e is a dimensionless constant ($e^2/4\pi = \alpha \approx 1/137$ in the $\hbar = c = 1$ system of units), A_α is a four-vector ($\alpha = 0, 1, 2, 3$) describing the electromagnetic field, and j_α is the electromagnetic current, which also is a four-vector. The dimension of A_α is that of mass $[A_\alpha] = [m]$, and the dimension of j_α is that of m^3. The expression $A_\alpha j^\alpha$ stands for the scalar product

$$A_\alpha j^\alpha = A_0 j_0 - A_1 j_1 - A_2 j_2 - A_3 j_3.$$

In quantum field theory, the electromagnetic current operator of the electron is expressed through ψ, the operator of annihilation of an electron and creation of a positron, and $\bar\psi$, the operator of creation of an electron and annihilation of a positron, as follows: $j_\alpha = \bar\psi \gamma_\alpha \psi$, where γ_α are four Dirac matrices. The electromagnetic currents of quarks contain the additional factors 2/3 or −1/3, which represent their fractional charges.

In the same way as electromagnetic currents of leptons and quarks are sources of the quanta of the electromagnetic field, namely, photons, so the weak currents of leptons and quarks are sources of the intermediate vector bosons W$^+$, W$^-$, and Z that mediate the weak interaction. The weak currents emitting and absorbing charged bosons W$^\pm$ are called charged currents; weak currents emitting and absorbing neutral bosons Z are called neutral currents. In contrast to the purely vector electromagnetic current, weak currents are a sum of a vector and an axial current. The interactions between weak currents are responsible for all known weak processes.

All strong processes are described by couplings of eight color currents to eight color gluons.

In grand unification models, there are currents responsible for the emission and absorption of superheavy X and Y bosons; these currents transform quarks into antiquarks or antileptons. The interactions of these currents produce proton decay.

Decay rate, or decay probability. A quantity characterizing the intensity of the decay of unstable particles; its dimension is s^{-1} and it is equal to the fraction of particles of an ensemble that decay per unit time. The decay rate is

$$w = -\frac{1}{N}\frac{dN}{dt} = \frac{1}{\tau},$$

where τ is the lifetime of a particle. The law of particle decay is exponential; thus, the fraction of particles surviving at time t is $\exp(-t/\tau)$. If the lifetime of a particle is sufficiently long, it can be measured by direct methods, for instance, by measuring the distance covered by a particle with known velocity from the point of creation to the point of decay. Lifetimes down to 10^{-16} s (lifetime of $\pi°$ mesons) are measurable in this way.

For particles with very small τ, the decay rate is measured by using the energy dependence of cross section given by the Breit-Wigner formula,

$$\sigma(E) = \sigma(M) \frac{(\Gamma/2)^2}{(E-M)^2 + (\Gamma/2)^2},$$

where Γ is the width of the resonance curve, $\sigma(M)$ is the cross section of the process at $E=M$, M is the mass of the particle, and E is the total energy of the decay products in the rest frame of the particle, for example, of two pions in the ρ-meson rest frame in the reaction

$$\pi^- p \to n\rho°$$
$$\hookrightarrow \pi^+ \pi^-$$

or the total energy of the initial particles, for example, of an electron and a positron in the reaction

$$e^+e^- \to J/\psi \to \pi^+ \pi^- \pi°$$

In the $\hbar = c = 1$ system of units we have used throughout this book, $\Gamma = w = 1/\tau$. Sometimes, instead of Γ, the total width Γ_{tot} is introduced to distinguish it from the partial widths Γ_i that characterize individual decay channels, $\Sigma_i \Gamma_i = \Gamma_{tot}$. A dimensionless quantity $B_i = \Gamma_i/\Gamma_{tot}$ is called the relative probability of a given channel, or branching ratio, $\Sigma_i B_i = 1$.

Deep inelastic processes. The processes of interaction between leptons and hadrons, accompanied by high transfer of energy E and momentum $\mathbf{q}$ (such that $|E^2 - \mathbf{q}^2| \gg 1$ GeV^2), resulting in multiple production of hadrons. The main deep inelastic processes are: (i) deep inelastic scattering of electrons, muons, and neutrinos on nucleons (often only these scattering processes are meant when deep inelastic processes are discussed), (ii) the annihilation of e^+e^- into hadrons at high energies in colliding beams, and (iii) the creation of lepton pairs (e^+e^- or $\mu^+\mu^-$) with large mass (i.e., with high total energy in the center-of-mass frame of the pair) in high-energy hadron collisions.

Processes similar to deep inelastic processes are those of hadron collisions, in which the so-called direct photons are produced, that is, photons with large transverse momenta p_T, and those in which hadron jets or individual hadrons are produced with large p_T. Such processes are said to be hard.

The first experiments on deep inelastic scattering of electrons on nucleons, carried out at Stanford (USA) in 1969, have revealed that a nucleon does not behave like a homogeneous jelly but contains pointlike constituents: hard grains that sharply change the momentum of the incident electron. In a certain sense, this discovery was analogous to the discovery of the atomic nucleus in Rutherford's experiments on the scattering of α particles.

Feynman christened the pointlike constituents of hadrons "partons." Subsequent experiments led to the conclusion that partons are quarks. In deep inelastic processes, quarks interact with leptons at short distances and, because of their asymptotic freedom, quarks behave like nearly free partons. The parton model explained the experimentally observed property of the scale invariance of deep inelastic scattering (Bjorken scaling, named after a well-known American theorist). Both the Bjorken scaling and the small deviations from it found in more accurate experiments are explained by quantum chromodynamics. According to quantum chromodynamics, not only quark partons but also gluon partons must participate in deep inelastic processes. This prediction is borne out by experimental data that show that gluon partons carry roughly one half of the total momentum of a fast hadron.

DESY (Deutsche Electronen Synchrotron). Acronym of a laboratory located near Hamburg (FRG) and of the accelerator installed there (German Electron Synchrotron), with a maximum electron energy of 7.5 GeV, which was put into operation in 1964. The cyclotron also feeds the electron-positron colliding beam facility DORIS. In the spring of 1978, the energy of each DORIS beam was raised to 5 GeV and, in summer of 1982, to 5.4 GeV; this made possible the observation of the resonance creation of upsilonium levels. Another facility in the DESY laboratory is the PETRA colliding beam accelerator.

Diagonal. This epithet implies that the corresponding expression is on the principal diagonal of a matrix. In the case of diagonal fermion currents, this is the matrix whose columns correspond to operators ψ_m, rows to operators $\bar{\psi}_n$, and each entry to a current j_{nm} of the type $\bar{\psi}_n \psi_m$. The diagonal current is that for which $m=n$; hence, this current transforms a particle m

into itself. If two currents are said to be coupled diagonally, the entries with $m=n$ in a similar matrix $j_m^+ j_n$ are meant. Diagonal currents and couplings do not change the flavors of the fundamental particles involved.

Dirac magnetic monopole. A hypothetical magnetic charge, not yet observed experimentally. The magnetic field produced by a monopole is similar to the field at the end of a sufficiently thin (infinitely thin) solenoid whose other end is at a sufficiently far remove (the field created by the second end of the solenoid then resembles the field of an antimonopole).

For an infinitely thin infinitely long solenoid string to be nonobservable, the flux of the magnetic field inside the solenoid must be quantized in such a manner that the phase of the wave function of an electron that circled the solenoid would be changed by $2\pi n$ (otherwise, an observable Bohm-Aharonov interference effect would be produced). This phase is $e \int \mathbf{A} \, d\mathbf{l} = e \int H° ds$, where $\mathbf{A}$ is the vector potential and $H°$ is the field inside the solenoid; the second integral is taken over the cross section of the solenoid. Taking into account the fact that the number of lines of force of the magnetic field is conserved, we find that the strength of a spherically symmetrical magnetic field H at a distance r from the monopole is $H = \int H° ds/4\pi r^2$. Expressing H in terms of the magnetic charge $(H=\mu/r^2)$, we obtain $4\pi\mu = 2\pi n/e$, or $\mu = n/2e$, where $e^2 = \alpha = 1/137$ (we again use the $\hbar = c = 1$ system of units).

In order to banish all possible confusion due to normalization, let us compare the potential between two electrons and the potential between two unit monopoles $(n=1)$. In $\hbar = c = 1$ units, the former is α/r, while the latter is $1/4\alpha r$.

New life was breathed into the theory of magnetic monopoles in 1974, when 't Hooft and Polyakov discovered that magnetic monopoles with finite mass must exist as classical solutions in a wide class of non-Abelian gauge models with spontaneously broken symmetry (e.g., in the SU(2) group with a triplet of Higgs fields). Such classical monopoles are predicted by grand unification models: SU(5), SO(10), etc. Their masses must be about 10^{16}–10^{17} GeV. Estimates show that the velocities of such massive monopoles coming to earth from space must be of the order of $10^{-3} c$. Because of this low velocity, the stopping power of matter for monopoles is expected to be very low; at present, we do not know how to slow down or stop them. But if we do learn how to stop them, it will be extremely interesting to observe the annihilation of a monopole and an antimonopole. Among the products of this annihilation must be the very heavy vector bosons (X and Y) which are responsible for proton decay in grand unification models.

There are no reliable estimates of the anticipated abundance of primordial monopoles left over from the Big Bang.

Double β decay. The β decay of an atomic nucleus, in which its charge Z changes by two and two electrons (or positrons) are emitted. In principle, two types of double β decay are possible: two-neutrino decay 2β (2ν) and neutrinoless decay 2β (0ν). Neither of the two types of double β-decay has been reliably detected in experiments.

Dyons. Hypothetical particles possessing both an electric and a magnetic charge; in other words, dyons are electrically charged magnetic monopoles. Dyons appear as solutions of equations in non-Abelian gauge theories.

Einstein-Podolsky-Rosen paradox. Reference to a thought (Gedanken) experiment discussed in a paper by these three authors published in 1935 under the title "Can a quantum-mechanical description of physical reality be considered complete?" In this thought experiment, two subsystems of a single quantum system fly away to a large distance from each other. But, however large the distance, the subsystems remain strictly correlated. Each subsystem is described not by an independent vector of state (ψ function), but by the so-called density matrix. As a result, a measurement of the state of one object instantaneously "reduces" the state of the second object.

The system of concepts and rules of quantum mechanics is internally consistent and supported by numerous experiments. There is nothing paradoxical in the framework of quantum mechanics in the instantaneous reduction, but quite a few physicists believe that the instantaneous reduction contradicts intuitive notions concerning causality, and discussion of the Einstein-Podolsky-Rosen experiment still continues in the literature. This discussion may ultimately lead to a more profound understanding of the process of measurement in quantum mechanics.

Electric dipole moment of a particle. One of the physical quantities characterizing the interaction of a particle with a static electromagnetic field (the other quantities are the electric charge, magnetic dipole moment, electric and magnetic quadrupole moments, etc.). The energy of the interaction between a dipole moment **d** and the electric field **E** is $-\mathbf{d}\cdot\mathbf{E}$. The vector **d** of such simple objects as elementary particles, atomic nuclei, and atoms can align, in contrast to more complicated objects, only along the spin **J**; hence, the product $\mathbf{d}\cdot\mathbf{E}$ is proportional to the product $\mathbf{J}\cdot\mathbf{E}$. Spatial inversion reverses the sign of **E** (**E** is a polar vector), but **J** (an axial vector) retains its sign; time reversal does not affect the sign of **E** but reverses the sign of **J**. Consequently, the coupling d**E** and, hence, the very existence of the electric dipole moment, are possible only if both the mirror invariance and the invariance under time reversal are violated in nature.

After CP violation and T violation effects were discovered in decays of K° mesons, it became obvious that elementary particles must have nonvanishing electric dipole moments. However, the value of the moment cannot be predicted until the mechanism of T violation is understood (and with it, by virtue of the CPT theorem, the mechanism of CP violation). The lowest experimental upper bound on the dipole moment is established for the neutron: $|d_n| \lesssim |e| \cdot 4 \times 10^{-25}$ cm, where e is the electron charge (compare this figure with the magnetic moment of the neutron, $|\mu_n| \simeq |e| \cdot 2 \times 10^{-14}$ cm).

Exotic baryons. Baryons whose quantum numbers are such that they cannot be constructed out of three quarks (qqq) but must contain at least four quarks and one antiquark (qqqq$\bar{\text{q}}$). Consider, for example, the so-called Z baryons, whose strangeness is +1, so that they include a valence antiquark $\bar{\text{s}}$. If Z baryons exist, they should manifest themselves as resonances in the scattering of K^+ mesons by nucleons. Some experimental groups reported observations of such resonances, but they were not included in the *Review of Particle Properties* by the Particle Data Group in 1982. Nonstrange baryon resonances with isotopic spin 5/2, whose existence is predicted by some theoretical models of hadrons, would be another example of exotic baryons.

Baryons with a structure qqqq$\bar{\text{q}}$ or qqqg, where g is a gluon, or even more complex, whose quantum numbers are identical to those of ordinary three-quark baryons, are called cryptoexotic (from the Greek word κρυπτός, "mysterious" or "hidden"). So far, the existence of cryptoexotic baryons has not been supported by experiment.

Exotic mesons. Mesons whose quantum numbers are such that they cannot be constructed of a quark and an antiquark (q$\bar{\text{q}}$) but must contain at least two quarks and two antiquarks (qq$\bar{\text{q}}\bar{\text{q}}$) or two quarks and a gluon (q$\bar{\text{q}}$g). Calculations in the framework of the bag model indicate that the state b$\bar{\text{b}}$g, with an exotic set of quantum numbers $J^{PC}=1^{-+}$ (i.e., a set forbidden in the quark + antiquark system), may be lighter than the Y''' meson.

Mesons with a structure more complex than q$\bar{\text{q}}$ but with quantum numbers identical to those of ordinary mesons are called cryptoexotic.

Fermions. Elementary or composite particles with half-integral spin. Fermions are governed by the Fermi-Dirac statistics. Only one fermion of a given type can occupy a given quantum state. This principle is called the Pauli principle (Wolfgang Pauli, 1900–1958; Pauli himself referred to it as the exclusion principle). The electron and other leptons, quarks, the proton and other baryons, atomic nuclei, and atoms with half-integral spin are fermions. The word fermion is derived from the name of the Italian physicist Fermi (Enrico Fermi, 1901–1954).

Feynman diagrams. Diagrams giving a graphic representation of interactions between particles. The main elements of Feynman diagrams (graphs) are lines representing the propagation of field perturbations (particles) and vertices representing their local interactions. Therefore, complex processes of interaction at a distance are reduced to elementary local interactions. Usually, the propagation of a fermion is shown by a straight line and that of a boson by a wavy line. Dashed, dash-dot, sawtooth, double lines, etc. are used when a process involves several sorts of particles. Feynman diagrams give a relativistically invariant description of processes. Correspondingly, the 4-momentum is conserved not only in the propagation of particles but also in the vertices.

Feynman diagrams are the basis of the relativistically invariant perturbation theory. In the calculations, each internal line is put into correspondence with the propagator of a particle and each vertex with the corresponding term of the interaction Lagrangian. Integration is carried out over 4-momenta of particles forming loops. Feynman diagrams thus define the algorithms for calculating the amplitudes of various processes.

We have mentioned several times that the notion of a quantized field is incomparably richer than that of a particle. This distinctive feature is manifested, for example, in that, for the general case of fields with spin greater than or equal to unity, one has to take not only the propagators of particles but also the so-called "ghosts" into account; this was noticed by Feynman, DeWitt, Faddeev, Popov, and Mandelstamm.

Feynman scaling characterizes the spectra of particles in processes of multiple production in collisions of high-energy hadrons; it states that the shape of the spectrum of produced particles is independent of the energy of the primary particle in the range of high energies of this primary. It is a function of only two variables: a dimensionless variable, x, equal to the longitudinal momentum of the particle produced, p_l, divided by the maximum energy of the spectrum and the transverse momentum of the particle, p_T. Some manifestations of this law were found in cosmic ray studies.

Later, beginning with the first experiments at Serpukhov at the end of the 1960s, this phenomenon was investigated on accelerators with much better accuracy. Then, also at the end of the 1960s, Feynman gave its analysis in terms of the parton model, and the phenomenon acquired the name "Feynman scaling."

FIAN. The Lebedev Institute of Physics of the USSR Academy of Sciences, Moscow.

Flavor. The characteristics of the type of quark (for a given color) or lepton. Such quantum numbers as strangeness, charm, and beauty are distinct quark flavors. There are six quark flavors, u, d, s, c, b, and t (the sixth quark, t, has not yet been observed), and six lepton flavors, e, μ, τ, ν_e, ν_μ, and ν_τ. Flavor is conserved in strong and electromagnetic interactions and is not conserved in weak interactions.

FNAL. The Fermi National Accelerator Laboratory (Batavia, near Chicago, USA) now known as Fermilab.

Functional. A numerical function on some linear space (space of functions). For instance, the area bounded by a closed curve of a given length is a functional.

Functional integral, or path integral. The limit of a multiple integral when the number of integrations tends to infinity.

Gauge symmetry. Invariance of the Lagrangian under some group of continuous transformations (Lie group) whose parameters are functions of space-time coordinates.

Examples of unbroken gauge symmetries are the Abelian group $U(1)_{em}$ describing the interaction of photons with charged particles and the non-Abelian group $SU(3)_c$ describing the color interactions of gluons with themselves and with quarks. Examples of spontaneously broken gauge symmetries are the group $SU(2)_W \times U(1)_Y$ of the standard model of electroweak interactions (subscript W for weak isospin and Y for weak hypercharge) and the groups of grand unification models (SU(5), SO(10), etc.).

A gauge symmetry requires the existence of massless gauge vector fields (photons, gluons, W and Z bosons, and X and Y bosons). Spontaneous breaking of a gauge symmetry gives masses to at least some of them.

The term gauge invariance (in German, Eichinvarianz) was introduced in 1919 by Weil, who used it in the same sense as scale invariance

(Maßstabinvarianz). Later, when quantum mechanics was born, gauge invariance was defined as the simultaneous transformation of the phase of the wave function of a charged particle,

$$\psi \to \psi' = \psi e^{i\alpha(x)},$$

and of the electromagnetic potential,

$$A_\mu \to A_\mu' = A_\mu + \partial_\mu \alpha$$

(V.A. Fock, 1926; F. London, 1927; H. Weil, 1929). Gauge transformations were often called gradient transformations in the Russian literature, but that term has been gradually disappearing in recent years.

Glueball (the same as gluonium). A colorless hadron (meson) consisting of two or more valence gluons but no valence quarks. The existence of glueballs is predicted by quantum chromodynamics. These particles must be looked for among the decay products of heavy quarkonia, such as J/ψ and Υ mesons. Two meson resonances were discovered in 1981 in radiative decays of the J/ψ meson: ι (iota) and θ (theta). The iota meson has a mass of 1440 MeV, zero spin, and negative parity; the theta meson has a mass of 1640 MeV, spin 2, and positive parity. Some authors believe that these mesons are glueballs, but this has not been proved.

Gluodynamics. A simplified quantum field theory of the strong interaction in which, as in quantum chromodynamics, there is an octet of colored gluons interacting with one another, but in which there are no quarks. Gluodynamics is studied in order to better understand some aspects of chromodynamics.

Gluon condensate. The nonzero vacuum expectation value of the operator $F^a_{\mu\nu} F^a_{\mu\nu}$, where $F^a_{\mu\nu}$ is the strength of the gluon field. In terms of this nonperturbative vacuum expectation value, the energy density of the gluon vacuum ϵ is given by

$$\epsilon = -\frac{9}{32} < 0 \mid \frac{\alpha s}{\pi} F^a_{\mu\nu} F^a_{\mu\nu} \mid 0 > \simeq -(\frac{1}{4}\text{GeV})^4.$$

This value of ϵ was determined by a theoretical analysis of the experimental data on charmonium and other mesons, in the framework of the so-called quantum-chromodynamical sum rules. The gluon condensate plays an important part in determining the physical properties of hadrons.

Gluons. Eight elementary massless spin 1 particles. Eight gluons form a color octet: The gluons of the octet differ only in color. The emission and absorption of gluons by quarks is the mechanism of the strong interaction between quarks. The theory of the interaction between gluons and quarks is called quantum chromodynamics. Since a gluon possesses a color charge, it can emit or absorb another gluon and thereby change color. As a result of this peculiar property of gluons, the effective color charges of gluons and quarks diminish as momentum transfer increases (as distance decreases), so that the strong interaction weakens (this is the so-called asymptotic freedom). As distance increases, the color interaction becomes stronger. This property may cause the confinement of quarks and gluons.

Gravitation (from the Latin "gravitas," "gravity"). The universal attraction between all particles, universal gravitation. Gravitational attraction is not proportional to the mass of a particle, as stated in high-school textbooks, but depends rather on its total energy and momentum. Thus, light or radio waves arriving from a distant star are deflected by the gravitational field of the Sun despite the zero mass of the photon. The characteristic energy at which the gravitational interaction becomes strong is the Planck mass $m_P = (\hbar c/G_N)^{1/2} \simeq 1.2 \times 10^{19}$ GeV $\simeq 10^{-5}$ g.

The masses of known elementary particles and the energies available in accelerators being negligible in comparison with m_P, the role played by the gravitational interaction in contemporary experimental high energy physics is correspondingly negligible. However, gravitation plays an important, and possibly even the key, role in fundamental theoretical physics and in the theory of elementary particles. Each year the number of theoretical papers devoted to gravitation and its possible links with other interactions steadily increases.

The classical (as opposed to the quantum) theory of gravitation, that is, general relativity, is a well-developed theory confirmed by a number of quantitative observations. General relativity is the basis of modern cosmology. The quantum theory of gravitation, quantum gravity, has not yet been constructed. Among the various approaches to the construction of quantum gravity, theoretical models of supergravity appear to be the most promising.

Gravitational constant (Newton constant), G_N. The constant characterizing the force of gravitational attraction. Two nonrelativistic particles with masses m_1 and m_2 at a distance r from one another are attracted by a force

$G_N m_1 m_2 r^{-2}$, where $G_N = 6.6720(41) \times 10^{-8}$ cm^{3}g^{-1}s^{-2} = $6.7065(41) \times 10^{-39} \hbar c^5$ GeV^{-2}.†

Graviton. The quantum of the gravitational field, a massless neutral spin-2 particle. Owing to the extreme weakness of the gravitational interaction, the experimental observation of gravitons is a challenge far beyond the capabilities of today's experimental physics.

Gravity waves. Oscillating gravitational fields emitted by bodies moving with variable accelerations, propagating in vacuum at the speed of light. To detect both the surges of gravitational radiation from extraterrestrial sources (e.g., from the collapsing nuclei of galaxies) and the gravitational waves generated in the laboratory, gravitational antennae were designed in about twenty laboratories of different countries. So far, these experiments have failed to give any positive results, presumably because of insufficient sensitivity.

Group. A set G of elements $g_1, g_2, \ldots$ equipped with a binary operation of composition, · (often referred to as multiplication), satisfying the following axioms.

1. There is an identity element e such that

$$g \cdot e = e \cdot g = g.$$

2. For each element g, there exists an inverse element g^{-1} such that

$$g^{-1} \cdot g = g \cdot g^{-1} = e.$$

3. The product of three elements is associative:

$$g_1 \cdot (g_2 \cdot g_3) = (g_1 \cdot g_2) \cdot g_3.$$

If all elements commute, that is, if $g_i \cdot g_k = g_k \cdot g_i$, the group is called commutative, or Abelian; otherwise, the group is non-Abelian.

A subgroup H of a group G is defined as a subset of elements of G satisfying conditions (1–3). A subgroup is said to be invariant if $g^{-1}hg \in H$ for any two elements $h \in H$ and $g \in G$. The identity element e and the group G itself are said to be trivial invariant subgroups of G.

A linear representation of a group G is defined as a mapping of the group G onto a group of linear transformations (of matrices) in some linear space,

---

†Here and throughout the Glossary, the number in parentheses indicates the uncertainty of one standard deviation in the last significant digits of the main figure: $6.6720(41) \equiv 6.6720 \pm 0.0041$.

such that the elements of this group are in a single-valued correspondence with the elements of G.

The groups important in physics are groups of transformations corresponding to various symmetries; they are referred to as symmetry groups. Especially important among them are the Lie groups.

Hadrons. Particles participating in strong interactions. Integral-spin hadrons are called mesons and half-integral-spin hadrons are called baryons. The number of known hadrons runs into the hundreds.

Most hadrons are extremely unstable: they are the so-called resonances which decay into lighter hadrons through the strong interaction. The lifetime of resonances is less than 10^{-21} s.

Quasistable hadrons live much longer and decay through the weak and electromagnetic interactions. The decay products of quasistable hadrons are leptons and photons, and lighter mesons. Decays of heavy mesons may also produce baryon-antibaryon pairs. The lightest baryons, the proton and neutron, are called nucleons. The heavier quasistable baryons (Λ, Σ, Ξ, Ω, Λ_c, . . .) are called hyperons. The decay products of hyperons are leptons, photons, mesons, and always a nucleon or a lighter hyperon.

Protons and neutrons are the building blocks of atomic nuclei. Other hadrons cannot be found in the stable matter that surrounds us: They are created in collisions of high-energy particles. The sources of such particles are accelerators and cosmic rays.

According to the current understanding, hadrons are not truly elementary particles: They are composed of quarks. The word hadron derives from the Greek word 'αδρος, for "massive, strong, large."

Higgs bosons. Hypothetical spin-zero particles that play an important role in the standard model of the electroweak interaction and in other theories that involve spontaneous symmetry breaking.

Hypercharge. A quantum number characterizing an isotopic multiplet. The hypercharge is equal to twice the mean electric charge (in $|e|$ units) of the particles that form the multiplet. This definition holds both for isotopic multiplets of hadrons and for isotopic multiplets of the helicity states of leptons and quarks in the gauge theory of the electroweak interaction. Examples: the hypercharge of the proton and neutron forming an isotopic doublet is $+1$ and that of the right-handed electron e^-_R, which is an isosinglet, is -2.

Identical particles. All elementary particles of a given type are identical. This is one of the most profound properties of elementary particles. The number of electrons in the universe is approximately 10^{80}; all these electrons are identical and indistinguishable—and the same is true for protons, and neutrons, and atoms built out of these particles. The unstable particles of each given type created in collisions at high energies are also identical to one another. Furthermore, all bosons of a given type exist in the universe in the state that is symmetrical with respect to their permutations, while all fermions of a given type are in the antisymmetrical state. These properties of bosons and fermions manifest themselves in the Bose-Einstein statistics for the former and the Fermi-Dirac statistics for the latter. A photon that has just been emitted is already symmetrized with all other photons in the universe; a new-born electron is already antisymmetrized to all other electrons. To provide for this property of bosons and fermions in quantum field theory, operators of boson creation commute and operators of fermion creation anticommute.

An objection can be raised against the picture outlined above: Experimentally, the indistinguishability of particles can be verified only within a certain accuracy, while the statement on their total identity is essentially absolute. Could different electrons be unlike one another "a teeny-weeny bit"? The answer to this question is as follows. The mathematical apparatus of today's theory rejects this "teeny-weeny" difference: the minutest difference discretely increases the number of degrees of freedom (the number of types) of particles and changes their statistics. Consequently, at the present state-of-the-art, we cannot phenomenologically parametrize the accuracy to which this indistinguishability is verified. This parametrization would necessitate a revolutionary restructuring of quantum field theory. So far I have not seen any convincing proposals for such a parametrization in the literature.

IHEP. Institute of High Energy Physics (Serpukhov, Protvino, Moscow Region). IHEP has a proton accelerator with a proton beam energy of 76 GeV (ring circumference 1.5 km). A project exists to begin construction of a proton-antiproton collider with a particle energy of 3 TeV in each beam (ring circumference 20.7 km).

Instanton. A special type of vacuum fluctuation of the gluon field; it belongs to the class of so-called nonperturbative phenomena, that is, phenomena not described by perturbation theory. Instanton-like vacuum fluctuations may play an important part in the mechanism of confinement of gluons and quarks.

In Minkowski space, instantons describe the quasiclassical trajectories of below-barrier transitions between topologically distinct vacuum states of gauge fields.

An interpretation of instantons not in the Minkowski space, but in the four-dimensional Euclidean space (with imaginary time) is more illustrative. Here instantons represent solutions of classical Yang-Mills equations with finite action.

Instanton solutions were discovered in 1975 by Belavin, Polyakov, Schwartz, and Tyupkin. In the case of the local group SU(2) in Euclidean space, the gauge field of an instanton located at the origin of the coordinate system is

$$A^a_\mu(x) = -\frac{2}{g} \frac{\eta_{a\mu\nu} x_\nu}{x^2 + \rho^2}.$$

Here a is an isovector index, $a=1, 2, 3$; μ, ν, and λ are indices of Euclidean coordinates, μ, ν, $\lambda = 0, 1, 2, 3$; and g is the running constant of the gauge interaction (invariant charge). The scale parameter ρ determines the instanton size. The dimensionless quantity $\eta_{a\mu\nu}$ is called " 't Hooft's symbol" (by the way, the term "instanton" was coined by 't Hooft),

$$\eta_{a00} = 0, \ \eta_{a0i} = -\eta_{ai0} = \delta_{ai}, \ \eta_{aij} = \epsilon_{aij},$$

where ϵ_{aij} is a completely antisymmetrical tensor ($a, i, j = 1, 2, 3$). The so-called anti-instanton solution is obtained from the instanton solution by the substitution $\eta_{aij} \to \bar\eta_{aij}$, where

$$\bar\eta_{a00} = 0, \ \bar\eta_{a0i} = -\bar\eta_{ai0} = -\delta_{ai}, \ \bar\eta_{aij} = \epsilon_{aij}.$$

The tensor of strength of an instanton gauge field is

$$F^a_{\mu\nu}(x) = -\frac{4}{g} \frac{\rho^2 \eta_{a\mu\nu}}{[x^2 + \rho^2]^2}$$

and the corresponding Euclidean action is

$$S^{(E)} = \frac{1}{4} \int F^a_{\mu\nu} F^a_{\mu\nu} d^4 x = \frac{8\pi^2}{g^2}.$$

The contribution of instantons to the amplitudes of physical processes is proportional to $\exp(-S^{(E)})$; it is very small for small-sized instantons (owing to the smallness of $g^2(\rho)/4\pi$ for ρ much less than the confinement radius). For larger instantons, the quasiclassical approximation breaks down because of large quantum corrections. Therefore, when calculating under the approximation of a "rarified gas of small instantons," theorists hope that the corresponding formulas, as often happens in physics, will be approximately valid far beyond the range in which they are definitely applicable.

Intermediate bosons (equivalent terms: intermediate vector bosons, weak vector bosons). Spin 1 particles (the charged W^+ and W^- and neutral Z) emitted and absorbed by weak currents and thus mediating the weak interactions of quarks and leptons. So far (through the end of 1982), intermediate bosons have not been observed experimentally. According to the standard model of the electroweak interaction, the masses of the W bosons must be about 80 GeV, and that of the Z boson about 90 GeV. The widths of their decays into the presently known particles must be around 2 GeV. The production and decay of intermediate bosons will be observable† at proton-antiproton colliders and at LEP and SLC. By measuring the total width of the Z boson it will be possible to establish the total number of light particles (including neutrinos) participating in weak neutral currents.

ITEP. Institute of Theoretical and Experimental Physics (Moscow).

ITP. L.D. Landau Institute of Theoretical Physics of the USSR Academy of Sciences (Chernogolovka).

JINR. Joint Institute of Nuclear Research (Dubna).

KNO scaling (KNO for Z.Koba, H.Nielsen, P.Olesen) characterizes the multiplicity distribution of events in multiple production of hadrons. Let us denote the multiplicity in a given event by n and the mean multiplicity at a given energy by $<n>$. The mean multiplicity of hadrons $<n>$ is known to increase with the energy of colliding hadrons. KNO scaling states that the multiplicity distribution of events at a given energy depends only on $n/<n>$ and is independent of the energy of colliding particles. The width of the multiplicity distribution thus increases proportionally to the mean multiplicity $<n>$ and not to $\sqrt{<n>}$, as we would expect for the Poisson distribution. Experimentally, the KNO scaling holds only approximately, having been verified up to the energy of the CERN $p\bar{p}$ collider.

---

*See Author's note to p. 113.

Lagrangian. A fundamental physical quantity; it occupies the central place in elementary particle physics and determines all properties of physical fields. The Lagrangian is usually denoted by the capital script letter $\mathscr{L}$. The Lagrangian determines the equations of propagation and interaction of fields by using the principle of least (to be precise, extremal) action. The action S is the integral of the Lagrangian $\mathscr{L}$ over space and time,

$$S = \int \mathscr{L} dx\, dy\, dz\, dt.$$

The integral

$$L = \int \mathscr{L}\, dx\, dy\, dz$$

is called the Lagrange function.

The construction of a theory of elementary particles is often said to be reducible to solving two problems: (i) finding the form of the Lagrangian of nature's fundamental fields and (ii) finding the experimentally verifiable corollaries of the Lagrangian of the given type.

The Lagrangian of quantum field theory is a sum of various terms, each of which contains a product of field operators or their derivatives. The Lagrangian of standard quantum field theories is local, that is, the fields entering the products refer to one and the same space-time point (the same values of $\mathbf{r}$ and t).

As was proved by E. Noether, the invariance of the Lagrangian under various groups of transformations implies the corresponding conservation laws. Ideally, a concrete form of the Lagrangian should be completely determined by symmetry principles, but the Lagrangians discussed in the literature actually contain a number of terms and parameters that are, so to speak, introduced "by hand," in order to describe the observable physical picture of the world. This is especially true for the scalar sector of the theory, that is, for those terms of the Lagrangian that include scalar fields.

Action is dimensionless in the $\hbar = c = 1$ system of units, $[S] = 1$, and the dimensionality of the Lagrangian is (mass)^4, $[\mathscr{L}] = [m^4]$. The terms in the Lagrangian can be divided into three groups: kinetic terms, mass terms, and coupling terms (in gauge theories, the kinetic terms and the terms describing the couplings of vector fields are closely linked). The dimension of the fermion field operators ψ in the Lagrangian is (mass)$^{3/2}$, $[\psi] = [m^{3/2}]$, and that of boson field operators is $[\varphi] = [m]$. In the general case, the coefficients in front of the different terms must be dimensional, in order to ensure that $[\mathscr{L}] = [m^4]$. The renormalizability of a Lagrangian necessitates that the dimensions of these coefficients be nonnegative, $[m^n]$, $n \geq 0$.

LEP (for Large Electron Positron (Ring)). A collider of electron-positron beams under construction at CERN. The ring circumference is about 27 km (with a tolerance of ±2 cm). The energy of each beam is 50 GeV, with a spread of about 80 MeV. The expected luminosity is of the order of 10^{31} cm^{-2}s^{-1}. This collider, whose estimated cost is approximately one billion Swiss francs, is planned to start operations at the end of 1988. The foremost task of the LEP is the study of the creation and decay of Z bosons. Since the cross section of Z creation in e^+e^- collisions ($e^+e^-\to Z$) is 4×10^{-32} cm^2, LEP is expected to produce one Z boson every 2–3 s. The energy of each beam will be subsequently raised to 80 GeV (stage II) and then to 125 GeV (stage III). This energy will make it possible to observe the creation of charged W$^\pm$ bosons: $e^+e^-\to W^+W^-$.

Leptons. Spin 1/2 elementary particles not participating in strong interactions. Three charged leptons are known—the electron e^-, muon μ^-, and τ^--lepton—and three neutral leptons— the electron neutrino ν_e, muon neutrino ν_μ, and tau neutrino ν_τ. Each of these particles has a corresponding antiparticle: e^+ (positron), μ^+, τ^+, $\bar\nu_e$, $\bar\nu_\mu$, and $\bar\nu_\tau$ (reads: "anti-nu-tau"). Electromagnetic interactions create pairs of charged leptons: e^+e^-, $\mu^+\mu^-$, and $\tau^+\tau^-$. In weak decays, each charged lepton is created in the company of its antineutrino: $e^-\bar\nu_e$, $\mu^-\bar\nu_\mu$, and $\tau^-\bar\nu_\tau$. If we assume that all leptons possess a special quantum number, the lepton number (sometimes called the lepton "charge"), equal to +1 and that all antileptons have a lepton number of −1, then the lepton number thus defined is conserved in all processes observed until now. The processes in which nonconservation of the lepton number is anticipated are proton decay, double β decay, and neutrino oscillations.

The muon and the τ-lepton are unstable, owing to the weak interaction. The electron is stable.

The word lepton derives from the Greek word λεπτός, "slender, narrow" (cf. lepta, a small Greek coin).

Lie algebra. A linear space L equipped with an operation [,] called commutation and having the following properties: (1) $[al_1 + bl_2, l_3] = a[l_1, l_3] + b[l_2, l_3]$ (linearity), (2) $[l_1, l_2] = -[l_2, l_1]$ (antisymmetry), (3) $[l_1[l_2, l_3]] + [l_2[l_3, l_1]] + [l_3[l_1, l_2]] = 0$ (Jacobi's identity) for any $l_1, l_2 \ldots \in L$. An algebra is said to be real (complex) if the numbers $a, b, \ldots$ are real (complex).

Example: the algebra of Pauli matrices τ_1, τ_2, and τ_3 with the commutator $[\tau_i, \tau_k] \equiv \tau_i\tau_k - \tau_k\tau_i$.

An algebra is called Abelian (commutative) if $[l_i, l_k] = 0$ for arbitrary l_i, $l_k \in L$. A subalgebra N is called an ideal if $[l,n] \in N$ for all $l \in L$ and $n \in N$. An algebra L is said to be simple if it is not Abelian and has no ideals distinct from $\{0\}$ and L. An algebra is called the direct sum of algebras M and N if $[m,n] = 0$ for any $m \in M$ and $n \in N$. An algebra is said to be semisimple if it is a direct sum of simple ideals.

The relation of a Lie group G to a Lie algebra L can be symbolically written in the form

$$G \sim \exp L.$$

Lie groups. Groups of continuous transformations whose elements are analytic functions of a finite number of continuous parameters; they bear the name of the Norwegian mathematician Sophus Lie (1842–1899).

Among Lie groups are the Poincaré group (a group of four-dimensional translations and rotations in a four-dimensional space), the unitary Abelian group $U(1)$, and the unitary unimodulus non-Abelian groups $SU(n)$, $n \geq 2$, which play an important role in the theories of strong and electroweak interactions and in grand unification models.

If the parameters of a group are independent of space-time coordinates, then the group and the corresponding symmetry are called global; otherwise, they are said to be local, or gauge, groups and symmetries.

A Lie group is said to be simple if it contains no nontrivial invariant subgroups with the possible exception of discrete subgroups. A Lie group is said to be semisimple if it contains no nontrivial invariant Abelian subgroups, with the possible exception of discrete subgroups.

The number of independent parameters on which the element of a Lie group depends is called the dimension of the group. A Lie group is said to be compact if its group manifold is compact.

A mapping of a group G on a group of matrices put in a single-valued correspondence with the elements of G is called the matrix representation of the group G. In the case of Lie groups, a special role is played by matrices realizing transformations arbitrarily close to the identity transformation: $G = 1 + d\omega_i I_i$, where $d\omega_i$ are infinitesimal parameters of the transformation and I_i are the so-called generators of a given representation. The number of linearly independent generators is equal to the dimension of the group. The maximum number of mutually commuting linearly independent generators is called the rank of the group.

The number of linearly independent vectors in the linear space in which matrices operate is called the dimension of the representation. (In the case of internal symmetry, the dimension of a representation is the number of particles in the corresponding multiplet.)

The simplest representations from which all other representations of a group can be constructed by means of multiplication are said to be fundamental (in the case of the groups SU(n), n-component spinors). The dimension of the adjoint representation is equal to the dimension of the group.

According to Cartan's classification, all compact simple Lie groups are subsumed into four regular series of groups—SU($l+1$), SO($2l+1$), Sp($2l$), and SO($2l$), corresponding to algebras A_l, B_l, C_l, and D_l, whose rank l can be arbitrarily large, $l=1, 2, \ldots$—and five exceptional groups—G_2, F_4, E_6, E_7, and E_8 (the subscript gives the rank of the group).

Let us list the basic groups (not necessarily compact simple or semisimple) of $n \times n$ matrices M (d is the dimension of the group)

GL(n,C) — *general* (G) *linear* (L) group of *complex* (C) *regular* (det $M \neq 0$) matrices; $d = 2n^2$

SL(n,C) — *special* (S: det $M = 1$) linear group, subgroup of GL(n,C); $d = 2(n^2 - 1)$

GL(n,R) — general linear group of *real* (R) regular matrices; $d = n^2$

SL(n,R) — special linear group of real matrices, a subgroup of GL(n,R); $d = n^2 - 1$

U(n) — unitary group of *unitary* (U: $MM^+ = M^+M = 1$, where M^+ is the Hermitian conjugate of M) matrices; $d = n^2$

SU(n) — special unitary group, a subgroup of U(n); $d = n^2 - 1$

O(n,C) — *orthogonal* (O) group of complex orthogonal matrices ($M\tilde{M} = 1$, where $\tilde{M}$ is the transposed M); $d = n(n-1)$

O(n)≡O(n,R) — orthogonal group of real orthogonal matrices, $d = n(n-1)/2$.

SO(n) — special orthogonal group or group of rotations in n-dimensional space, a subgroup of O(n); $d = n(n-1)/2$

Sp(n) — *symplectic* (Sp) group of unitary n×n matrices, where n is even, satisfying the condition $\tilde{M}JM = J$, where J is a nonsingular antisymmetrical matrix.

U(m,$n-m$) — pseudounitary group of complex matrices satisfying the condition $MgM^+ = g$, where g is a diagonal matrix with elements $g_{kk} = 1$ for $1 \leq k \leq m$ and $g_{kk} = -1$ for $m+1 \leq k \leq n$; $d = n^2$

O(n,$n-m$) — pseudo-orthogonal group of real matrices satisfying the condition $MgM = g$; $d = n(n-1)/2$.

Luminosity. The number of collisions per second per unit cross section; characterizes colliding-beam facilities, the so-called colliders. Luminosity is usually denoted by L; its dimension is cm^{-2}s^{-1}. Luminosity times the cross section of a given process, σ, in cm^2 gives the number of the relevant events per second.

Majorana neutrino. A truly neutral neutrino, a neutral spin 1/2 particle, transforming into itself under charge conjugation. The theory of such a truly neutral neutrino was proffered by the Italian physicist Ettore Majorana (1906–1938). If neutrinos are massless and always emitted with "Weyl helicity," then the Majorana neutrinos are indistinguishable from the ordinary two-component neutrinos (the so-called Weyl neutrinos, whose polarization is left-handed for neutrinos and right-handed for antineutrinos). However, if the neutrino mass is distinct from zero, then the theory of Majorana neutrinos will predict a number of specific effects, among them the double β decay.

Majoron. A hypothetical neutral spin 0 massless (or very light) particle very weakly interacting with matter. The existence of the majoron is predicted by some theoretical models in which the neutrino obtains a Majorana mass as a result of spontaneous breaking of the conservation of the leptonic quantum number.

Mass. A relativistically invariant quantity characterizing a particle or a system of particles. The relation of the mass m of a body to its energy E and momentum $\mathbf{p}$ is

$$m^2 c^4 = E^2 - \mathbf{p}^2 c^2,$$

where c is the speed of light. In the relativistic system of units where $c=1$, $m^2 = E^2 - \mathbf{p}^2$.

Sometimes the mass m is referred to as the "rest mass," in order to distinguish it from the "motion mass" defined as E/c^2. Both these terms are obsolete. They are a vestige of the faraway time when special relativity was coming of age and nonrelativistic formulas were sometimes used to describe relativistic particles and, in particular, the Newtonian relation between momentum and velocity, $\vec{\mathbf{p}} = m\vec{\mathbf{v}}$, instead of the relativistic formula $\mathbf{p} = E\mathbf{v}/c^2$. Nowadays, the term "rest mass" is hardly ever encountered in serious books on physics but is quite frequent on the pages of science-popularizing publications. To label both the Lorentz invariant quantity and the component of the Lorentz vector by the same noun "mass" is extremely unfortunate. Furthermore, there is no logic in calling the same quantity, energy, in this case, by two different names: Whatever the units of measurement, energy is always energy. A term like "motion mass" would be especially absurd in the system of units in which $c=1$.

As for the gravitational attraction, it is proportional to the energy-momentum tensor of a particle, not to its mass. Gravitational force acts not on the mass but on energy and momentum; this is why massless photons bend their paths in gravitational fields.

Mesons. Hadrons with integral spin. All mesons have zero baryon quantum number.

The word meson derives from the Greek word μεσος, "intermediate." When the term was proposed, it was implied that mesons' masses were intermediate: greater than the electron's but less than the proton's. At the present time, this shade of meaning has faded away because some of the known mesons are considerably more massive than the proton.

It is utterly wrong to attribute the name "μ meson" to the muon: the muon is a lepton, not a meson. This misplaced term, still cropping up in the current literature, was generally accepted in the 1930s and 1940s, when the modern classification of elementary particles had not yet evolved. Neither is it correct to apply the term "meson" to the intermediate vector bosons W and Z or to scalar (Higgs) bosons: neither of them are hadrons.

Multiplet. An ensemble of particles or states (energy levels) of a system having similar properties.

Multiplet of hadrons. An ensemble of hadrons with identical spin and parity, close values of mass, and similar strong interactions. Multiplets appear because of the symmetries of the strong interaction. The transformations of the appropriate symmetry group convert particles of the multiplet into one another. The first hadron multiplet, namely, the doublet of nucleons, was introduced into physics by Heisenberg in the 1930s when the neutron was discovered. The proton and the neutron differ in their electromagnetic properties (charge, magnetic moment, and the distribution of charge and current inside the particle). As for the rest, these two particles are very similar: They have identical spin, nearly identical mass (a difference of the order of 0.1 percent), and practically identical strong interactions. When interactions of nucleons are studied, it is natural to ignore (in the first approximation) the difference between the proton and the neutron and to treat them as two degenerate states of one particle, the nucleon, assuming that nuclear forces are invariant under transformations of these states into each other. A mathematical description of the nucleon doublet is analogous to a description of a spin 1/2 particle, with its two-component spinors, Pauli matrices, and the rest of the "equipment" of the SU(2) group.

The proton-neutron symmetry was called the isotopic symmetry. Here the term "isotopic" is used not in the sense of the symmetry of nuclear isotopes: In the terminology of nuclear physics, the proton and neutron are not isotopes but isobars. An attempt was even made in this connection to replace the term "isotopic symmetry" with "isobaric symmetry." However, the latter term proved uncompetitive.

A nucleon is described by a spinor in the isotopic space. The proton and the neutron correspond to projections of the isotopic spin onto some axis (the z axis) in the isotopic space: projections $+1/2$ and $-1/2$, respectively.

The next isotopic multiplet, the triplet of pions, was discovered at the end of the 1940s and the beginning of the 1950s. In the 1950s and 1960s, numerous isotopic multiplets of strange particles and resonances were discovered. The first isotopic multiplets of charmed particles were found in the 1970s. The number of particles in an isotopic multiplet, n, is related to the isotopic spin I by the simple formula $n = 2I + 1$.

The explanation of the isotopic symmetry was supplied by quantum chromodynamics. That is, u and d quarks are practically interchangeable because the difference between their masses is small compared to the characteristic energies of u and d quarks within hadrons.

If the s quark were as light as the u and d quarks, all three quarks would be interchangeable, and the SU(3) symmetry corresponding to this interchangeability would be just as good as the isotopic SU(2) symmetry. In nature, the SU(3) symmetry is violated much more strongly than SU(2). The reason for this is the rather large mass of the s quark:

$$m_s - m_u \approx m_s - m_d \approx 150 \text{ MeV}.$$

The simplest SU(3) multiplets of hadrons are singlets, octets, and decuplets.

Higher "flavor" symmetries of hadrons, SU(4), SU(5), . . . are evidently almost completely broken in nature because the masses of heavy quarks c, b, . . . are much greater than their characteristic momenta within hadrons.

Neutral currents. Weak currents determining the couplings of leptons and quarks to neutral intermediate bosons Z. The interactions of neutral currents mediated by virtual Z bosons result in a number of specific effects, for example, the muonless neutrino reactions discovered in 1973 and the parity nonconservation in electron-nucleon interactions discovered in 1978.

All known neutral currents conserve the flavor of the participating leptons and quarks and are diagonal, that is, contain the operators of both creation and annihilation of the same particle, for instance, $\bar{\nu}_\mu \nu_\mu$ and $\bar{e}e$.

Occam's razor. A principle stating that "Entities ought not to be multiplied except out of necessity" (Entia non sunt multiplicanda praeter necessitatem"). This principle was formulated by the English philosopher William Occam, or Ockham (1285–1349).

Oscillations of K mesons (from Latin "oscillare," to swing). The mutual transformations of $K°$ and $\bar{K}°$ mesons in vacuum observed in beams of neutral K mesons. These transformations result from the weak interactions between the quarks of which K mesons consist: $K° = \bar{s}d \leftrightarrow s\bar{d} = \bar{K}°$. Owing to these transformations, the states $K°$ and $\bar{K}°$ do not have definite masses and definite lifetimes. The states with definite masses and lifetimes are the $K°_S$ and $K°_L$ mesons. The first of these is a short-lived meson (subscript S for short), with a lifetime $\tau_S \simeq 0.9 \times 10^{-10}$ s; the second is long-lived (subscript L for long), with $\tau_L \simeq 5.2 \times 10^{-8}$ s. The $K°_L$ meson is slightly heavier than the $K°_S$ meson: $m_L - m_S \simeq 3.5 \times 10^{-6}$ eV $= 0.53 \times 10^{10}$ s^{-1}. The period of $K° \leftrightarrow \bar{K}°$ oscillations is $\tau = 2\pi/(m_L - m_S) \simeq 1.2 \times 10^{-9}$ s. Oscillations of K mesons were first observed experimentally at the end of the 1950s and have now been very thoroughly investigated.

Oscillations of neutrinos. Mutual transformations of different types of neutrinos and antineutrinos (ν_e, ν_μ, ν_τ, $\bar{\nu}_e$, $\bar{\nu}_\mu$, and $\bar{\nu}_\tau$) in the vacuum. The putative neutrino oscillations have been discussed in the literature since the middle 1950s, but an intensive experimental search began only in recent years. So far, the search for neutrino oscillations with accelerators, reactors, and in cosmic rays has given no evidence that these oscillations exist. The fact that the observed flux of solar neutrinos is roughly three times less than the expected value is regarded as an indirect argument in favor of the neutrino oscillation hypothesis. That is, it is conjectured that, on the way from the Sun to the Earth, the electron neutrino, ν_e, is converted into a mixture of equal amounts of ν_e, ν_μ, and ν_τ (the last two sorts of neutrinos appear to be "sterile" at low energies; i.e. they cannot produce the reactions $\nu + Cl \rightarrow e^- + Ar$, which is used to detect solar neutrinos).

It is necessary (but not sufficient) for the existence of neutrino oscillations that the neutrinos have nonzero masses.

Oscillations of neutrons. Hypothetical mutual transformations of neutrons and antineutrons in vacuum. These transformations require that there be an interaction capable of changing the baryon quantum number B by two, because $B(n)=+1$ and $B(\tilde{n})=-1$. An interaction with this property appears in some grand unification models. Note that the interaction resulting in the decay $p \to e^+\pi^\circ$ cannot produce the transition n ↔ ñ because the former conserves the difference between the baryon and lepton quantum numbers, $B-L$, while this difference changes by two in the latter. Therefore, the experiments searching for the neutron-antineutron oscillations are supplementary to the experiments looking for proton decay. The n ↔ ñ oscillations in vacuum are very sensitive to the strength of the hypothetical interaction with $\Delta B = 2$ because the neutron and antineutron masses are identical and even a superweak interaction would be sufficient to produce mixing of degenerate levels. By using intense beams of neutrons from nuclear reactors and high current accelerators (up to 10^{17} neutrons s^{-1}), it is possible to detect n ↔ ñ transitions if the oscillation period τ_{osc} is less than or of the order of 10^{10} s. The fraction of antineutrons in the beam increases quadratically with the time t in which the neutrons cover the distance from the source to the detector: $N_{\tilde{n}}/N_n \propto t^2/\tau_{osc}^2$.

In nuclei, n ↔ ñ transitions will be observed as transformations of two nucleons of the nucleus into mesons. However, these transitions are strongly suppressed in comparison with n ↔ ñ transitions in vacuum because the energy level of the antineutron in a nucleus is substantially separated from that of the neutron and has a very large annihilation width.

Parity. The quantum number that characterizes the symmetry of the wave function of a particle or system of particles under a discrete transformation. In the case of P parity (spatial parity), the transformation is the mirror reflection with respect to three mutually orthogonal planes intersecting at the origin of coordinates. In the case of C parity (charge conjugation parity), the transformation is the charge conjugation: the replacement of particles by the corresponding antiparticles. The CP parity (or combined parity, in Landau's phrase) is the product of P and C parities. In 1956–1957, it was discovered that weak processes are not invariant under both mirror reflection and charge conjugation. The effects of P and C violation in weak processes are large, close to unity. In 1964, very small effects of CP violation were observed in decays of long-lived neutral K mesons.

The nonconservation of P parity stems from the fact that a weak current is the sum of a vector and an axial vector. Mirror reflection reverses the sign of the vector but leaves that of the axial vector unaltered. The vector and axial vector currents are also transformed differently under charge conjugation. As for the CP violation, its source is so far unknown.

In the case of bosons, the P parity of a particle and that of its antiparticle are identical and their product is +1. The product is −1 for fermions. According to the standard convention, the parity of the Dirac fermion is postulated to be +1; thus, the parity of the antifermion is −1. When truly neutral fermions (e.g., the Majorana neutrino) are considered, the P parity of fermions and antifermions has to be made identical and, therefore, imaginary, i or $-i$.

Parity violating nuclear forces. Weak interactions between nucleons revealed in mirror-asymmetric effects in atomic nuclei. Examples of such effects are the P-odd angular distribution and the circular polarization of nuclear γ quanta, the P-odd angular distribution of fission fragments of uranium and thorium, and the circular polarization of photons in the reaction n + p → d + γ, among others.

PEP (Proton-Electron-Positron (Storage Ring)). Electron-positron storage ring at SLAC (Stanford Linear Accelerator Center). Tunnel length 2.2 km; energy of each beam 18 GeV; luminosity 3×10^{30} $s^{-1}cm^{-2}$. PEP became operational in September 1980. The proton mentioned in the acronym PEP is a vestige of one of the originally discussed uses of the facility for the study of ep collisions.

PETRA (Positron-Electron Tandem Ring Accelerator). Colliding electron-positron beams in the DESY laboratory near Hamburg, FRG. The accelerator has been in operation since April 1979. The ring length is 2.3 km. The maximum energy attained in 1981 was 2×19 GeV; the maximum luminosity $L = 1.7 \times 10^{31}$ $cm^{-2}s^{-1}$.

Phenomenology. In current theoretical physics, the classification and description of phenomena (empirical data) based on known laws applied to the external attributes of phenomena with no clarification of their profound nature or internal mechanism. The term derives from the Greek φαινομενον, "appearing."

Photon. An elementary particle with zero mass and a spin 1. The photon has no charge, being a truly neutral particle. Depending on their energy, photons fall into the ranges of radio waves, visible light, x rays, or hard γ quanta. The emission and absorption of photons by charged particles is the basis of all electromagnetic processes.

Planck mass, m_P. The mass given by the formula

$$m_P = (\hbar c/G_N)^{1/2},$$

where G_N is the gravitational constant; $m_P = 1.221(4) \times 10^{19}$ GeV/$c^2 \simeq 2.18 \times 10^{-5}$ g.

Planck's constant, $\hbar$. The quantum of action. $\hbar = h/2\pi = 1.0545887(57) \times (10^{-27}$ erg s $= 10^{-34}$ J s$) = 6.582173(17) \times 10^{-25}$ GeV s.

Pomeranchuk theorem. The theorem stating that the cross sections of interaction of a particle and its antiparticle (e.g., a proton and antiproton) with the same target tend to the same limit as energy increases. (I.Ya.Pomeranchuk, 1913–1966).

Positronium. An atom-like system consisting of an electron and a positron bound by the Coulomb attraction. Depending on the value of the orbital momentum L, the levels of positronium are denoted by the capital letters S, P, D, F, G ... for $L=0, 1, 2, 3, 4$ ..., respectively. Depending on the value of the toal spin S of the electron and the positron, the levels of positronium are classified into singlet levels ($S=0$, the so-called parapositronium) and triplet levels ($S=1$, orthopositronium). The ground states of para- and orthopositronium are $^1S_0$ and $^3S_1$, respectively. The superscript here indicates the spin multiplicity of the level, $2S+1$, and the subscript gives the total angular momentum of the level, J.

The spatial parity of a level is $P=(-1)^{L+1}$ and its charge conjugation parity is $C=(-1)^{L+S}$. By virtue of the conservation of angular momentum and charge parity, the ground state of positronium decays into two photons, and that of orthopositronium into three photons. The levels of quarkonia, that is, systems consisting of a quark and an antiquark of the same flavor, such as charmonium, are classified by analogy with positronium levels.

Probability amplitude. A quantity in quantum mechanics and quantum field theory whose squared modulus determines the probability of the process. In the case of the collision of two particles, the probability of the process is characterized by the process cross section; in the case of the decay of an unstable particle, it is characterized by the decay width. Both the cross section and the decay width are proportional to the squared modulus of the amplitude.

Quantum (German "quant," Latin "quantum": how much). The term has several different meanings:

1. Quantum of field: a particle constituting an elementary excitation of a given field, for example, the photon, or light quantum (γ quantum), is the excitation of the electromagnetic field and electrons and positrons are quanta of the electron-positron field.

2. Quantum of energy: the amount of energy carried away by a particle (e.g., a photon) when a system (e.g., an atom) jumps from one energy level to another. The energy of the photon, E, is related to its frequency, ω, by the famous equation $E = \hbar\omega$.

3. Quantum of action: the universal world constant $\hbar = 1.0545887(57) \times 10^{-27}$ erg s, called Planck's constant; $\hbar$ plays a fundamental role in quantum mechanics.

Quantum chromodynamics (QCD). The quantum theory of gluon and quark fields and their interactions due to color charges (from the Greek $\chi\rho\bar{\omega}\mu\alpha$, "color").

Quantum electrodynamics (QED). The quantum theory of electromagnetic (photon) and electron-positron fields and their interactions.

Quantum field theory. The theory of relativistic quantum phenomena. Quantum field theory is essentially the most fundamental of physical theories. Nonrelativistic quantum mechanics and relativistic classical field theory are its limiting cases: the first at velocities much less than the speed of light; the second at values of action much greater than $\hbar$. The fundamental idea of quantum field theory is that all particles are quanta of the corresponding physical fields. Quantum field theory studies the processes of the creation, interaction, and annihilation of elementary particles.

The methods of quantum field theory form the basis of quantum electrodynamics, the standard model of the electroweak interaction, quantum chromodynamics, and grand unification models.

Quantum mechanics. A fundamental physical theory describing the widest range of phenomena—elementary particles, atoms, molecules, and multiatomic systems—in which the characteristic action S_{char} is comparable to the quantum of action $\hbar$. Processes in which $S_{\text{char}} \gg \hbar$ are called quasiclassical. Finally, if $\hbar$ can be totally neglected, classical mechanics becomes valid.

Quantum-mechanical systems possess both wave and corpuscular properties. Quantum mechanics states that a number of questions can, in principle, be answered only in a probabilistic sense. Thus, the cross section of the collision of two particles and the lifetime of an unstable particle are essentially probabilistic.

In quantum mechanics, the so-called pure state of a particle or a system of particles is represented by a vector in the Hilbert space of states. A mixed state is described in the space of states as a density matrix.

Each observable physical quantity (energy, momentum, and angular momentum) is assigned a corresponding operator.

Quantum-mechanical states somewhat resemble the martians in Ray Bradbury's *Martian Chronicles*: In each state, each operator finds what it's looking for. If the state is the eigenstate of a given operator, the corresponding physical quantity has a definite eigenvalue (a definite quantum number). In this case, the result of applying an operator to a state vector reduces to multiplying the state vector by the eigenvalue of the operator. If the state is not an eigenstate of a given operator, its state vector can be expressed as a linear superposition of eigenvectors with different possible eigenvalues of the operator. The coefficients of this superposition are probability amplitudes; their squared magnitudes give the probabilities of specific values of the physical quantity considered.

Accordingly, there are two classes of quantum-mechanical problems: (i) the calculation of the eigenvalues of physical quantities, for instance, atomic and molecular energy levels; and (ii) the calculation of the probabilities of various processes.

If two operators commute (i.e., if the order in which they act on a state does not affect the result), then there exist complete sets of states that are eigenstates of both operators simultaneously. And if the operators do not commute, then, in the general case, they have no common eigenstates. Thus, a particle cannot simultaneously have a definite momentum p_x and a definite coordinate x; Heisenberg's uncertainty relation states that

$$\Delta x \cdot \Delta p_x \geq \hbar/2.$$

Quark condensate. The nonzero vacuum expectation value of the operator $\bar{\psi}\psi$, where ψ is the operator of the annihilation of a quark and creation of an antiquark and $\bar{\psi}$ is the operator of the creation of a quark and annihilation of an antiquark. The notation for the quark condensate is $\langle 0|\bar{\psi}\psi|0\rangle$. It is an example of nonperturbative effects. This effect violates the chiral invariance of QCD. A theoretical analysis of experimental data on processes involving mesons shows that, for light quarks,

$$\langle 0|\bar{\psi}\psi|0\rangle \simeq -(\tfrac{1}{4}\text{ GeV})^3.$$

Quarks. Spin 1/2 particles, constituents of hadrons. Ordinary (not exotic) baryons consist of three quarks; ordinary mesons consist of a quark and an antiquark. Five sorts (flavors) of quarks are known; three of them, d, s, and b, have an electric charge of $-1/3$, and two, u and c, a charge of $+2/3$. Hypothetically, there also exists a sixth quark, t, with an electric charge of $+2/3$. No hadrons containing a t quark have yet been observed in accelerator experiments because of the high mass of the t quark. The energies or luminosities of particle beams required for the creation of t quarks are not yet attainable.

According to quantum chromodynamics, strong interactions between quarks are caused by specific color charges of quarks. Quarks of each flavor exist in three distinct color species: "yellow," "blue," and "red." A quark of one color can transform into a differently "colored" quark by emitting a colored gluon. Quarks interact by exchanging gluons. The color states of quarks within hadrons are such that the total color charge of a hadron is zero. Hadrons are thus said to be colorless or white.

Although a group at Stanford University has reported observations of free fractionally charged quarks for a number of years, experiments of other groups searching for free quarks give negative results and most physicists are quite skeptical about the idea of free quarks. There is a hypothesis of confinement in quantum chromodynamics (its viability has not yet been proved), which states that colored particles (quarks and gluons and their colored combinations) cannot, in principle, exist in the free state. At the same time, the existence of quarks in hadrons is an experimentally established fact. The first indirect evidence of the existence of quarks was supplied by the classification of hadrons. Later direct collisions of leptons with individual quarks were detected in experiments on deep inelastic interactions between leptons and hadrons. These collisions take place deep inside the hadron and take a very short time, during which a quark fails to exchange a gluon with another quark and thus interacts as an almost free particle. The greater the momentum transfer, that is, the shorter the distance at which the lepton-quark collision occurs, the freer the quark's behavior. This property, a corollary of asymptotic freedom, signifies that quarks are neither quasiparticles nor collective excitations of hadronic matter, but truly elementary particles, like leptons. A possible nonelementarity of quarks and leptons could be revealed only by a still deeper penetration into these particles, that is, at still greater momentum transfers.

The term "quark" was introduced in 1964 by Gell-Mann, who borrowed it from James Joyce's novel *Finnegans Wake* (the protagonist of the story hears the gulls crying "Three quarks for Muster Mark" in his night dream).

In German, "quark" means "curd." The quark symbols u,d,s,c,b, and t stand for up, down, strange, charm, bottom (beauty), and top (truth).

Scaling. The scale invariance (self-similarity) of physical processes.

Singlet. The simplest multiplet, consisting of a single state or single particle. Two particles form a singlet spin state if their total spin is zero.

SLC (SLAC Linear Collider). An electron-positron linear collider planned for construction at the Stanford Linear Accelerator Center. Both the electron and the positron beams are to be accelerated in the same accelerator up to 50 GeV. The top view of the collider resembles a tennis racket with the linac as its handle. The two beams emerging from the linac separate and go along two arcs, to collide at the apex of the racket frame.

Solar neutrinos. Neutrinos produced in nuclear reactions inside the Sun. The main source of solar neutrinos are nuclear reactions of the hydrogen cycle in which four protons ultimately transform into a ^{4}He nucleus, two positrons, and two neutrinos. The hydrogen cycle consists of the following stages:

(1) Burning of protons.
    99.75%: $p + p \to d + e^+ + \nu_e$ $E_\nu^{max} = 0.42$ MeV
    0.25%: $p + e^- + p \to d + \nu_e$ $E_\nu = 1.44$ MeV

(2) Burning of d.
    $d + p \to {}^3\text{He} + \gamma$ $Q = 5.5$ MeV

(3) Burning of ^{3}He.
    86%: ${}^3\text{He} + {}^3\text{He} \to {}^4\text{He} + 2p$ $Q = 12.9$ MeV
    14%: ${}^3\text{He} + {}^4\text{He} \to {}^7\text{Be} + \gamma$ $Q = 1.59$ MeV

(4) Burning of ^{7}Be.
    90%: ${}^7\text{Be} + e^- \to {}^7\text{Li} + \nu_e$ $E_\nu = 0.861$ MeV
    10%: ${}^7\text{Be} + e^- \to {}^7\text{Li}^* + \nu_e$ $E_\nu = 0.383$ MeV
    0.015%: ${}^7\text{Be} + p \to {}^8\text{B} + \gamma$ $Q = 0.133$ MeV

(5) Burning of ^{7}Li.
    ${}^7\text{Li} + p \to {}^4\text{He} + {}^4\text{He}$ $Q = 17.3$ MeV

(6) Decay of ^{8}B.
    ${}^8\text{B} \to {}^8\text{Be} + e^+ + \nu_e$ $E_\nu^{max} = 14.06$ MeV

(7) Decay of ^{8}Be.
    ${}^8\text{Be} \to {}^4\text{He} + {}^4\text{He}$ $Q = 0.92$ MeV

Notation: E_ν is the energy of neutrinos in reactions with two-particle final states, E_ν^{max} is the maximum energy of neutrinos in three-particle final states, and Q is the total energy release in a reaction or decay. The percentage gives the yield of the relevant reaction.

The above list shows that the neutrinos emitted in decays of ^{8}B have maximum energy. Until recently, experimental searches were conducted only for these, the so-called "boron neutrinos." They were recorded in the experiment of R.Davies (USA) in a mine at a depth of 1.6 km. The neutrino detector was a tank filled with perchloroethylene (C_2Cl_4); the reaction used was suggested by B.Pontecorvo:

$$\nu_e + {}^{37}Cl \rightarrow {}^{37}Ar + e^-, \quad {}^{37}Ar \rightarrow {}^{37}Cl + e^+ + \nu_e$$

(the half-life of ^{37}Ar is $T_{1/2} = 35$ days). Over the period from 1970 to 1980, roughly two neutrinos per day were recorded in 600 tonnes of C_2Cl_4, which corresponds to 1.9 ± 0.3 SNU (1 SNU is the solar neutrino unit, equal to 10^{-36} reactions of neutrino capture per atom of the target per second). This number must be compared with the theoretical prediction, 7.6 ± 3.3 SNU. The disagreement of the theory and the experiment (both errors correspond to three standard deviations) may be caused by such factors as the detailed chemical structure of the Sun, turbulent flows cooling the central part of the Sun where the boron neutrinos originate, or an insufficiently reliable estimate of the percentage of the reactions ^{3}He + ^{3}He $\rightarrow$ ^{4}He + 2p and ^{3}He + ^{4}He $\rightarrow$ ^{7}Be + γ. There is also an alternative hypothesis that, on the way from the Sun to the Earth, neutrino oscillations convert approximately two-thirds of electron neutrinos into muon and tau neutrinos, which are practically unobservable at such low energies.

In contrast to boron neutrinos, the flux of proton neutrinos, from reactions p + p $\rightarrow$ d + e^+ + ν_e and p + e^- + p $\rightarrow$ d + ν_e, is predicted quite reliably because the cross sections of these reactions are less dependent on temperature, owing to their low Coulomb barrier. Proton neutrinos have low energies, so detecting them calls for a low-threshold detector. The gallium isotope ^{71}Ga can be used for this purpose, its neutrino detection threshold being 0.231 MeV. It is planned to install at Baksan laboratory in Caucasus a detector with 50 tonnes of gallium, which is expected to record one proton neutrino per day. If this prediction is confirmed, it will be necessary to conclude that the observed deficiency of boron neutrinos is not caused by neutrino oscillations.

Special relativity. A fundamental theory of the space-time properties of all physical processes.

Special relativity is based on two principles. The first of them states that no experiment conducted in a closed physical system can determine whether the system is at rest or moves with constant speed in a straight line (with respect to a system of infinitely removed stars). This principle is called the Galileo-Einstein relativity principle; the corresponding frames of reference are said to be inertial. The second principle states that there is a limiting velocity (universal constant c) of propagation of physical objects and signals. (The velocity c is the velocity of the propagation of photons ("light") and other massless particles in the vacuum.) Consequently, all physical phenomena, including the propagation of light (and hence, all laws of nature) manifest themselves identically in all inertial frames of reference. This invariance of the laws of nature is known as the Lorentz invariance.

It is readily verified that the Lorentz invariance of the laws of nature holds only if the spatial and temporal distances between events (world points) change from one reference frame to another. Thus, spatially separated events that are simultaneous in one reference frame are not simultaneous in other frames. Therefore, the notion of simultaneity loses the absolute sense that it has in Newtonian mechanics.

The transformations of the relative space-time coordinates (x,y,z,t) of two events a and b ($x = x_a - x_b$, $y = y_a - y_b$, $z = z_a - z_b$, $t = t_a - t_b$) to a new inertial reference frame moving at a velocity $\mathbf{v}$ with respect to the original reference frame are called Lorentz transformations; in the case where $\mathbf{v}$ is parallel to the x axis, they have the form

$$x' = \frac{x - vt}{\sqrt{1 - v^2/c^2}}, \, y' = y, \, z' = z, \, t' = \frac{t - vx/c^2}{\sqrt{1 - v^2/c^2}}.$$

It is not difficult to verify that these transformations leave the interval $c^2 t^2 - x^2 - y^2 - z^2$ unchanged (invariant).

The coordinates x^0, x^1, x^2, x^3 (ct, x, y, z) can be regarded as the coordinates of a four-dimensional vector in Minkowski space. Lorentz transformations correspond to pseudorotations in this space. The energy E and momentum $\mathbf{p}c$ of a particle (or system of particles) also form a four-dimensional vector. The mass of a particle (or a system of particles) m is an invariant, $m^2 c^4 = E^2 - \mathbf{p}^2 c^2$.

The generators of three Lorentz transformations along space coordinate axes, together with the generators of three spatial rotations, form the Lorentz algebra and the associated Lorentz group. It can be shown that the Lorentz group is a single-valued (but not one-to-one) mapping of the group SL(2,C). If we add the generators of four space-time translations to the generators of the Lorentz group, we obtain the Poincaré algebra and Poincaré group.

There are two types of Lorentz vectors: contravariant vectors $x^\mu = x^0, x^1, x^2, x^3$ and covariant vectors $x_\mu = x_0, x_1, x_2, x_3$. The relation between them is

$$x_\mu = \eta_{\mu\nu} x^\nu, \quad x^\mu = \eta^{\mu\nu} x_\nu,$$

where $\eta_{\mu\nu}$ is the so-called metric tensor and the repeating indices (so-called dummy indices) imply summation. In the metric tensor, only diagonal elements are nonzero,

$$\eta_{00} = -\eta_{11} = -\eta_{22} = -\eta_{33} = 1.$$

Sometimes the following notation is used:

$$\eta_{\mu\nu} = \mathrm{diag}\,(1, -1, -1, -1).$$

The scalar product of two vectors, u_μ and v_μ, is formed by using the metric tensor,

$$uv = u^\mu v_\mu = u_\nu v^\nu = u_\mu v_\nu \eta^{\mu\nu} = u^\mu v^\nu \eta_{\mu\nu} = u^0 v^0 - u^1 v^1 - u^2 v^2 - u^3 v^3 = u^0 v^0 - \mathbf{uv}.$$

Spontaneous symmetry breaking. A breaking of symmetry in which the Lagrangian has a certain symmetry but stable physical states described by the Lagrangian, including the vacuum, do not have that symmetry. In this situation, the symmetric states are unstable and a spontaneous (from Latin "sponte," "of its own (free) will") transition to nonsymmetrical stable states occurs under the action of infinitesimal perturbations. Examples: A needle set vertically on its tip falls down and a pea inside a beer bottle carefully balanced on the top of the bottle's bottom rolls down, spontaneously violating the cylindrical symmetry.

In quantum field theory, a spontaneous symmetry breaking can be realized by a scalar field φ in which the energy of self-action (i.e., the energy of a nonlinear interaction with itself) has the form $\lambda^2 (\varphi^2 - \eta^2)^2$, where λ is a dimensionless parameter and the dimensionality of the parameter η is that of mass. In this situation, the self-action energy is minimal (this corresponds to a stable vacuum) when the field φ is everywhere equal to η and not to zero, as would be the case if the self-action had the form $\lambda^2 (\varphi^2 + \eta^2)^2$.

The theory of unified electroweak interaction introduces an isotopic doublet of scalar fields φ with gauge coupling to four massless vector fields. As a result of a spontaneous breaking of the symmetry $SU(2) \times U(1)$ to $U(1)_{em}$, the field φ obtains a nonzero vacuum expectation value η; the vector bosons W^+, W^-, and Z obtain masses of the order of $e\eta$, where e is the electric charge; and only the photon remains massless. This phenomenon is known as the Higgs mechanism. The original isotopic doublet of scalar fields then reduces to a single neutral field, that of the so-called Higgs boson. So far it has not been proved that precisely this model is realized in nature because Higgs bosons have not yet been observed experimentally. A search for these bosons is one of the most important tasks in high energy physics.

Supermultiplet. A supersymmetric multiplet comprising both fermions and bosons. The simplest supermultiplet contains one vector and one spinor particle, for instance, the photon and the massless neutral photino. This supermultiplet is called the gauge multiplet of $N = 1$ supersymmetry. The $N = 1$ supersymmetry also has another type of supermultiplets, namely, the chiral supermultiplet comprising a massless Majorana neutrino and two spin 0 neutral massless bosons, one scalar and one pseudoscalar. Variants of extended supersymmetry ($2 \leq N \leq 8$) have supermultiplets with a larger number of particles.

Symmetry (from the Greek σύμμετρον, "jointly measured"). The property of an object or ensemble of objects of retaining their shape or relative positions under certain transformations. The notion of symmetry is inseparable from the notion of beauty. Furthermore, true beauty, in its highest forms, requires a slight departure from symmetry, imparting to beauty the mysterious and alluring element of incompleteness.

In fundamental physics, symmetries are usually classified into geometrical and internal symmetries. Transformations corresponding to geometrical symmetries include spatial and temporal translations, spatial rotations, space-time rotations, and mirror reflections of coordinate axes (three spatial and one temporal). Except for the last of these, the symmetry under each one corresponds to the conservation of the appropriate quantity: momentum, energy, engular momentum, Lorentz momentum, and spatial parity. Symmetry under reversal of the time axis corresponds to the reversibility of physical processes.

Transformations corresponding to internal symmetries usually relate different particles "of like kind". Thus, charge conjugation transforms particles into the corresponding antiparticles, isotopic transformations transform different components of isotopic multiplets into one another, color transformations transform different components of color multiplets into one another, and so on. Correspondingly, these symmetries imply conservation of charge conjugation parity, isotopic spin, color, and so forth.

The geometrical and internal symmetries are not totally isolated from each other. Thus, the charge conjugation C, mirror reflection P, and time reversal T are related by the CPT theorem, so that CP violation entails violation of time reversibility. Another example is given by supersymmetry, which relates bosons and fermions. Two consecutive transformations of supersymmetry contain a space-time translation.

Symmetry transformations form groups. If transformations commute, the corresponding symmetry is called Abelian, otherwise it is said to be non-Abelian.

Vacuum, physical. The lowest energy state of a system of quantized fields; all physical processes occur against this background. Owing to quantum effects (the creation of pairs of virtual particles and the more complex vacuum fluctuations of quantized fields), the physical vacuum has a complicated structure; its description may involve nonzero quantum numbers. Sometimes the vacuum in the Newtonian sense, or in the sense of perturbation theory, is called the mathematical vacuum.

VEPP. The Electron-positron colliding beam accelerator at the Institute of Nuclear Physics of the Siberian Branch of the USSR Academy of Sciences (Novosibirsk). The energy of each beam of VEPP-2M is 0.7 GeV, and that of VEPP-4 is close to 5.5 GeV.

Virtual particles. In terms of Feynman diagrams, these are particles for which the condition $E^2 - \mathbf{p}^2 = m^2$ does not hold, in contrast to real, free particles; here E is the energy of a particle, $\mathbf{p}$ its momentum, and m its mass. To emphasize the violation of the equality $E^2 - \mathbf{p}^2 = m^2$, virtual particles are said to lie off the mass shell. The emission and absorption of virtual particles form the basis of practically all physical processes.

The word "virtual" derives from the Latin "virtus," "strength, virtue"; among the many meanings of the word, the most relevant in this context are "possible," "not real" (cf. the term "virtual displacements" in mechanics).

Yang-Mills fields. Vector fields representing gauge fields of a non-Abelian gauge symmetry. Example: gluon fields as gauge fields of the color group $SU(3)_c$. Non-Abelian massless gauge fields (for the group SU(2)) were considered for the first time in 1954 by Yang Chen Ning and R.L.Mills.

Yukawa coupling. The emission or absorption of a spin 0 boson by a spin 1/2 particle. The dimensionless coefficient characterizing the intensity of this coupling is called the Yukawa coupling constant. Examples: the emission or absorption of a π meson by a nucleon or a Higgs boson by a quark or lepton. (Hideki Yukawa (1907–1981), the Japanese theoretical physicist, predicted the existence of the π meson in 1935.)

In the general case, the terms of the Lagrangian describing the Yukawa coupling have the form

$$\bar{\psi}_b (f + f' \gamma_5) \psi_a \varphi,$$

where ψ_a is the annihilation operator for a spin 1/2 particle a and the creation operator for the antiparticle $\bar{a}$; $\bar{\psi}_b = \psi^+_b \gamma_0$, where $^+$ denotes Hermitian conjugation; and φ is the operator of scalar (pseudoscalar) field, $\gamma_5 = i\gamma_0\gamma_1\gamma_2\gamma_3$. If P parity is conserved, one of the Yukawa coupling constants (f or f', depending on the parity of the field φ) is zero.

Appendix 3

BIBLIOGRAPHY

On papers and preprints
On conferences and schools
Introductory comments to the list of reviews
Sections and subsections of the list of reviews
List of reviews

ON PAPERS AND PREPRINTS

The literature on elementary particle physics is vast and continues to grow rapidly. Thus, the bibliographical index *Particles and Fields*, published by the USSR Institute of Scientific and Technical Information (VINITI), reported the publication of about 6000 papers in 60 journals in 1971 and about 7500 papers in more than 100 journals in 1981. About 1500 papers, rapporteur talks, and lectures published in 1981 in the Proceedings of conferences and schools must be added to this figure.

In recent years, roughly one-half of all papers, reports, and lectures have first appeared as preprints published by various institutions. SLAC issues a weekly list of these preprints, *Preprints in Particles and Fields*. This source reported about 5000 preprints in 1982.

Similar lists of preprints are published by a number of other organizations, such as DESY and ICTP (the International Center for Theoretical Physics, in Trieste). CERN and JINR regularly publish lists of preprints, periodicals, and books received by the libraries of these research organizations.

Among the journals publishing original papers (not reviews) on elementary particle physics, the leading ones are: *Physical Review D, Nuclear*

Physics, Physical Review Letters, Physics Letters B, Zeitschrift für Physik C, and *Lettere al Nuovo Cimento*. In the USSR, the leading journals in this field are *Yadernaya Fizika, Zhurnal Eksperimental'noi i Teoreticheskoi Fiziki (ZhETF), Pis'ma v ZhETF,* and *Teoreticheskaya i Matematicheskaya Fizika*.†

ON CONFERENCES AND SCHOOLS

The most representative among international conferences in the field are the Rochester Conferences on High-Energy Physics, which assemble about one thousand participants every two years. The last five conferences were:
   London (1974; XVII), Tbilisi (1976; XVIII), Tokyo (1978; XIX, Madison (1980; XX), and Paris (1982; XXI).

In the years skipped by Rochester conferences, somewhat lesser-scale European Conferences on High-Energy Physics are held: Palermo (1975; VIII), Budapest (1977; IX), Geneva (1979; X), Lisbon (1981; XI), and Brighton (1983; XII).

International Photon and Lepton Conferences are held in the same years: Stanford (1975; VII), Hamburg (1977; VIII), Batavia (1979; IX), Bonn (1981; X), and Ithaca (1983; XI).

There are also large International "Neutrino 19 . . " Conferences: Blatonfüred (1975; V), Aachen (1976; VI), Baksan Valley (1977; VII), West Lafayette (1978; VIII), Bergen (1979; IX), Erice (1980; X), Honolulu (1981; XI), and Budapest (1982; XII).

In addition to these main conferences, about one hundred specialized conferences, symposia, and schools are organized annually, with smaller numbers of participants. Let us mention the recently published Proceedings of some traditionally convened annual symposia and schools:

   XVII International School of Subnuclear Physics of the "Ettore Majorana" Center for Scientific Culture. Erice 1979. Edited by A. Zichichi. New York, London, Plenum Press (1982). (The 20th School took place in 1982.)

   Proceedings of XVII Rencontre de Moriond. March 1982. Edited by Tran Thanh Van. Éditions Frontières, 1982.

   Proceedings of Orbis Scientiae. University of Miami, Coral Gables, v.8. Chairman B. Kursunoglu. New York, London, Plenum Press (1981).

   Proceedings of the 1981 CERN-JINR School of Physics. Hanko, Fin-

---

†*Translater's Note:* These journals are translated and published in English by the American Institute of Physics under the titles *Soviet Journal of Nuclear Physics, Soviet Physics JETP, JETP letters,* and *Theoretical and Mathematical Physics,* respectively.

land, 1981. Geneva 1982. (These are the Proceedings of the 7th CERN School; in odd years the school is conducted jointly with JINR.)

Fundamental Interactions. Cargèse 1981, 22nd Cargèse Summer Institute, Edited by M.Levy, et al. NATO Advanced Study Institute Series. Series B. Physics. Volume B85. New York, Plenum Press (1982).

Proceedings of the XVIII International Universitätswochen für Kernphysik 1979 at Schladming, February–March 1979. Edited by P.Urban. Acta Physica Austriaca Supplementum XXI. Vienna, New York, Springer Verlag, (1979).

Proceedings of the 9th SLAC Summer Institute on Particle Physics. July–August 1981. The Strong Interaction. SLAC 0245 (1981).

Elementary Particles. The 9th ITEP School of Physics, 1981, issues 1–3. Moscow, Energoizdat (1982) (in Russian).

Proceedings of XVII LINP† Winter School, 1981. High Energy Physics. Leningrad (1982) (in Russian).

XV International School for Young Scientists in High Energy Physics. Dubna (1981) (in Russian).

Proceedings of V International Seminar on Problems of High Energy Physics and Quantum Field Theory. Protvino, 1982, v.1, IHEP (1982).

The following regular, but not annual, seminars that were recently held in the USSR should be mentioned:

VI International Seminar on Problems of Quantum Field Theory. Alushta (1981).

VI International Seminar on Problems of High Energy Physics. Dubna (1981).

II International Seminar on Quantum Gravity. Moscow (1981). Quantum Gravity, edited by M.A.Markov and P.Wert, Plenum, 1983.

II International Seminar on Group-Theoretical Methods in Physics. Zvenigorod (1982).

Quarks-82 International Seminar. Sukhumi (1982).

The Divisions of General Physics and Astronomy and of Nuclear Physics of the USSR Academy of Sciences regularly convene, several times a year, joint sessions at which review talks, some of them on particle physics, are presented. The abstracts of these talks are published in *Uspekhi Fizicheskikh Nauk (Soviet Physics Uspekhi)*.

Unfortunately, the proceedings of even large international conferences are not always accessible to specialists, let alone students. This is all the more regrettable because the rapporteur talks at such conferences usually

---

†Leningrad Institute of Nuclear Physics.

present a comprehensive picture of the latest achievements in the most active fields of elementary particle physics, formulate unsolved problems, and outline approaches to finding their solutions.

INTRODUCTORY COMMENTS TO THE LIST OF REVIEWS

After a lengthy consideration of how to compile the bibliographic list for this book, I decided not to include in it the proceedings of conferences and schools because of their restricted availability, and to give instead a list of reviews published in the leading science-popularizing and review journals. The basis for my decision was the hope that at least some of these journals are accessible even to readers who live far away from main libraries.

The list of reviews comprises about 600 science-popularizing articles and review papers that appeared in 1976–1982 in *Scientific American (Sci.Am.)*, *American Journal of Physics (Am.J.Phys.)*, *Physics Today (Phys.Today)*, *Comments on Nuclear and Particle Physics (CONAPP)*, *Reviews of Modern Physics (Rev.Mod.Phys.)*, *Physics Reports (Phys.Rep.)*, *Uspekhi Fizicheskikh Nauk (UFN)*, and *Fizika Elementarnykh Chastits i Atomnogo Yadra (EChAYa)* (the last two journals are translated and published by the American Institute of Physics under the titles *Soviet Physics Uspekhi (Sov.Phys.Uspekhi)*, and *Soviet Journal of Particles and Nuclei (Sov.J.Part.Nucl.)*.†

Bibliographic descriptions of papers are classified into the sections and subsections listed at the end of this introductory text. References are grouped in each subsection within each journal title, in reversed chronological order: the more recent the publication date, the higher the paper in the list; journal-title groups follow in the order given in the preceding paragraph.

Although many review papers cover such a broad range of topics that they deserve to be placed in several subsections, I cite them only once in the list, where the reference seemed the most justified. This made the list more compact, even if less informative. This shortcoming is probably not too serious, as the contents of an article can often be guessed on the basis of its title.

The reader will be helped to some extent by cross references of the type "see also . . . " accompanying the headings of some sections and subsec-

---

†*Translator's Note*: The list of journals is abridged as compared to the Russian edition of this book by omitting references to *Priroda (Nature)*. This journal, which is published by the USSR Academy of Sciences, is not translated into English.

tions and indicating the other sections and subsections in which reviews relevant to a given topic can be found. (Note that there are no cross references among subsections of the same section.)

I did not try to stick to a list of section headings compiled beforehand, but instead grouped papers similar in content, assigning group headings later. As a result, the hierarchical structure of headings is broken in some places. I have sometimes assembled under a given subsection heading papers devoted to one particular aspect of one of the preceding subsections. This was done if the topic was represented by a large number of papers (see, for example, subsections 9.1 "General Relativity" and 9.2 "Black Holes").

In section 2 ("Particles"), only some particles were assigned individual subsections, namely, muons and neutrons (used in various applications-oriented studies), as well as neutrinos and the so-called new particles (discovered within the past decade), because they were subjects of numerous reviews. On the other hand, electrons and protons were not singled out in special subsections, so that proton decay, for example, must be looked for in section 7 ("Grand Unification").

Reviews on elementary particle physics were taken from the above eight journals without any bias. Papers are included even if I do not share the views of the authors (actually, such cases are not very frequent). My premise was that a complete list, when not regarded as a recommendation, is more useful to a reader than a subjectively truncated one.

I also tried to reflect in the list, at least partially, review papers in related fields: astrophysics, statistical physics, various applications of accelerators, and so on.

A cursory reading of the bibliographical list reveals a nonuniform distribution of papers among sections and subsections. Most of the reviews cited are devoted to various aspects of the physics of strong interactions. But many hot topics of principal importance (e.g., Higgs bosons and supersymmetry) are represented by a scanty four or five reviews. These topics are given a detailed treatment in lectures delivered at various schools and in the original papers cited in the reviews.

Extremely useful reference lists on selected subjects can be found in the so-called "Resource Letters" published time and again by the *American Journal of Physics* (e.g., see subsections 2.5, 5.1, and 9.2). Also worthy of attention are the summaries of science news regularly appearing in *Priroda*† under the heading "News of Science," in *Scientific American* under

---

†*See* footnote to p. 182.

the heading "Science and the Citizen," and especially in *Physics Today* under the heading "Search and Discovery." Only the fear of inflating the reference list to an unmanageable size stopped me from including references to these anonymous or semi-anonymous microreviews. Likewise, I refrained from adding to the reference list brief pedagogical notes from the *American Journal of Physics*.

Reviews in the list are very different in scientific level. Those appearing in science-popularizing *Priroda* and *Scientific American* are aimed at non-specialists. Papers in the *American Journal of Physics, Physics Today*, and *Comments on Nuclear and Particle Physics* are usually more complicated. As for those in *Soviet Physics Uspekhi, Physics Reports, Reviews of Modern Physics,* and *Soviet Journal of Particles and Nuclei,* they are mostly addressed to specialists. The uninitiated will be able to understand, at best, only a few introductory pages.

To the best of my knowledge, nobody has taken the trouble of compiling a bibliographic list similar to the one given below. Other potential authors had probably been stopped by the defects, apparent even to the naked eye, inherent in a bibliography of this sort. These defects are apparent to me as well. Nevertheless, I still believe that this list will prove useful to the reader.

SECTIONS AND SUBSECTIONS OF THE LIST OF REVIEWS

1. Experimental Techniques
    1.1. Accelerators (see also 3.2)
    1.2. Detectors

2. Particles
    2.1. Tables of particles
    2.2. Muons (see also 3.1)
    2.3. Neutrinos (see also 1.2, 5, 10)
    2.4. Neutrons (see also 5.2)
    2.5. New particles (see also 3.2, 4.2, 4.3, 4.5, 5.5)

3. The Electromagnetic Interaction
    3.1. Quantum electrodynamics
    3.2. e^+e^- collisions (see also 1.1, 2.5, 5.5)
    3.3. Electromagnetic interactions of hadrons

4. The Strong Interaction
  4.1. Quantum chromodynamics (see also 11.4, 11.10)
  4.2. Gluons
  4.3. Quarkonia (see also 2.5, 3.2)
  4.4. Bag model
  4.5. Free quarks
  4.6. Baryonium. Dibaryons. Hadronic atoms
  4.7. Processes involving a small number of particles
  4.8. Multiple production
  4.9. Large momentum transfer. Jets
  4.10. High-energy interactions in cosmic rays
  4.11. High-energy interactions of hadrons with nuclei
  4.12. Nuclear matter
  4.13. Semi-phenomenological approaches to strong interactions

5. The Weak Interaction
  5.1. Weak processe (see also 2.3, 2.5)
  5.2. CP violation (see also 2.4)
  5.3. Neutral currents
  5.4. Double β-decay
  5.5 Electroweak theory (see also 1.1, 3.2, 11.4)

6. Scalar Bosons and Technicolor (see also 11.4)

7. Grand Unification (see also 10.3, 11.4)

8. Supersymmetry and Supergravity

9. Gravitation (see also 8, 10, 11)
  9.1. General relativity
  9.2. Black holes

10. Astrophysics and Cosmology
  10.1 Astrophysics
  10.2. Primary cosmic rays
  10.3. Cosmology (see also 2.3, 7, 9)

11. Quantum Mechanics and Field Theory
  11.1. Quantum mechanics
  11.2. Path integrals

11.3. Symmetries
11.4. Gauge theories (see also 3.1, 4.1, 5.5, 7)
11.5. Classical solutions of gauge equations
11.6. Summation of diagrams. Renormalization group
11.7. Topology and differential geometry in physics
11.8. Nonlinear equations and solitons
11.9. Problems common for quantum field theory and statistical physics
11.10. Computers and field theory. Gauge fields on lattices (see also 4.1)
11.11. Stochastic behavior of dynamic systems
11.12. Axiomatic quantum theory

12. History and Prospects

LIST OF REVIEWS

1. Experimental Techniques

1.1. Accelerators (see also 3.2)

Wilson R.R. : The next generation of particle accelerators. *Sci. Am.* **242**, no.1 (1980) 26.

Paramentola J. and Tsipsis K. : Particle-beam weapons. *Sci. Am.* **240**, no.4 (1979) 38.

Yonas G. : Fusion power with particle beams. *Sci. Am.* **239**, no.5 (1978) 40.

Rowe E.M. and Weaver J.H. : The uses of synchrotron radiation. *Sci. Am.* **236**, no.6 (1977) 32.

Wilson R.R. : US particle accelerators at age 50. *Phys. Today* **34** (1981) 86.

Rowe E.M. : Synchrotron radiation: facilities in the United States. *Phys. Today* **34** (1981) 28.

Cline D. and Rubbia C. : Antiproton–proton colliders, intermediate bosons. *Phys. Today* **33** (1980) 44.

Panofsky W.K.H. : Needs versus means in high-energy physics. *Phys. Today* **33** (1980) 24.

Wilson R.R. : The Tevatron. *Phys. Today* **30** (1977) 23.

Muller R.A. : Radioisotope dating with accelerators. *Phys. Today* **32** (1979) 23.

Wilson R.R. : Fantasies of future Fermilab facilities. *Rev. Mod. Phys.* **51** (1979) 259.

Hahn H., Month M. and Rau R.R. : Proton–proton intersection storage accelerator facility ISABELLE at the Brookhaven National Laboratory. *Rev. Mod. Phys.* **49** (1977) 625.

Skrinskii A.N. : Accelerator and detector prospects of elementary particle physics. *Usp. Fiz. Nauk* **138** (1982) 3.†

Kapchinskii I.M. : Intense linear ion accelerators. *Sov. Phys. Uspekhi* **23** (1980) 835.

Yarba V.A. : Particle accelerators being constructed and planned at superhigh energies. *Sov. Phys. Uspekhi* **22** (1979) 841.

Budker G.I. and Skrinskii A.N. : Electron cooling and new possibilities in elementary particle physics. *Sov. Phys. Uspekhi* **21** (1978) 277.

Skrinskii A.N. and Parkhomchuk V.V. : Melthods of cooling beams of charged particles. *Sov. J. Part. Nucl.* **12** (1981) 223.

Balbekov V.I. et al. : The accelerator-storageing complex (UNK) at the Institute of High Energy Physics. *Sov. J. Part. Nucl.* **10** (1979) 222.

1.2. Detectors

Learned J.G. and Eichler D.A. : A deep-sea neutrino telescope. Sci. Am. **224**, no.2 (1981) 104.

Nygren D.R. and Marx J.N. : The time projection chamber. *Phys. Today* **31**, (1978) 46.

Sendweiss J. : The high-resolution streamer chamber. *Phys. Today* **31** (1978) 40.

Willis W.J. : The large spectrometers. *Phys Today* **31** (1978) 32.

Charpak G. : Multiwire and drift proportional chambers. *Phys. Today* **31** (1978) 23.

Pehl R.N. : Germanium gamma-ray detectors. *Phys. Today* **30** (1977) 50.

---

†Russian editions of *Soviet Phys. Uspekhi* and *Sov. J. Part. Nucl.* are cited when the corresponding 1982 issues in English were not available to the translator.

Charpak G. : Wire chambers: a review and forecast. *CONAPP* **6** (1976) 157.

Kleinknecht K. : Particle detectors. *Phys. Rep.* **84** (1982) 85.

Bellini G. , Foa G.M. , Sandweiss J. , Montanet L. , Reucrott S. and Prentice J. : Lifetime measurements in the 10^{-13} range. *Phys. Rep.* **83** (1982) 1.

2. Particles

2.1. Tables of particles

Particle Data Group: Review of particle properties. *Phys. Lett.* **111B** (April, 1982)

2.2. Muons (see also 3.1)

Nemethy P. and Highes V.W. : Additive versus multiplicative muon conservation. CONAPP **10** (1981) 147.

Schacher J. : Is there no muon-electron conversion? *CONAPP* **8** (1978) 97.

Karlson E. : The use of positive muons in metal physics. *Phys. Rep.* **82** (1982) 271.

Scheck F. : Muon physics. *Phys. Rep.* **44** (1978) 187.

Gurevich I.I. et al. : Physics and chemistry of the μ^+ meson and muonium. *Sov. J. Part. Nucl.* **8** (1977) 46.

2.3. Neutrino (see also 1.2, 10)

Mann A.K. : Neutrino oscillations. *CONAPP* **10** (1981) 155.

Marciano W.J. : Neutrino masses: theory and experiment. *CONAPP* **9** (1981) 169.

Bilenky S.M. and Pontecorvo B. : Lepton mixing and the Solar neutrino puzzle. *CONAPP* **7** (1977) 149.

Cline D. : Neutrino microscope, flavorscope and telescope. *CONAPP* **7** (1977) 121.

Bahcall J.N. : Solar neutrino experiments. *Rev. Mod. Phys.* **50** (1978) 881.

Frampton P.H. and Vogel P. : Massive neutrinos. *Phys. Rep.* **82** (1982) 339.

Bilenky S.M. and Pontecorvo B. : Lepton mixing and neutrino oscillations. *Phys. Rep.* **41** (1978) 225.

Aliev T.M. and Vysotskii M.I. : Prospects for detecting photons produced by the decay of primordial neutrinos in the universe. *Sov. Phys. Uspekhi* **24** (1981) 1008.
Bilen'kii S.M. and Pontecorvo B.M. : Lepton mixing and neutrino oscillations. *Sov. Phys. Uspekhi* **20** (1977) 776.

Borovoi A.A. : Reactor neutrino experiments. *Sov. J. Part. Nucl.* **11** (1980) 35.

2.4. Neutrons (see also 5.2)

Greenberger D.M. and Overhauser A.W. : The role of gravity in quantum theory. *Sci. Am.* **242**, no.5 (1980) 544.
Golub R. et al. : Ultracold neutrons. *Sci. Am.* **240**, no.6 (1979) 106.

Greenberger D.M. and Overhauser A.M. : Coherence effects in neutron diffraction and gravity experiments. *Rev. Mod. Phys.* **51** (1979) 43.

Lushchikov V.I. : Ultracold neutrons. *Phys. Today* **30** (1977) 42.

Sears V.F. : Fundamental aspects of neutron optics. *Phys. Rep.* **82** (1982) 1.
Ramsey N.F. : Dipole moments and spin rotations of the neutron. *Phys. Rep.* **43** (1978) 409.

Frank I.M. : Fiftieth anniversary of the discovery of the neutron. *Sov. Phys. Uspekhi* **25** (1982) 279.
Frank A.I. : Fundamental properties of the neutron: fifty years of research. *Sov. Phys. Uspekhi* **25** (1982) 280.
Ostanevich Yu.M. and Serdyuk I.N. : Neutron-diffraction studies of the structure of biological macromolecules. *Sov. Phys. Uspekhi* **25** (1982) 340.

2.5. New particles (see also 3.2, 4.2, 4.3, 4.5)

Lederman L. : Upsilon particle. *Sci. Am.* **239**, no.4 (1978) 60.
Perl M.L. and Kirk W.T. : Heavy leptons. *Sci. Am.* **238** no.3 (1978) 50.

Schwitters R.F. : Fundamental particles with charm. *Sci. Am.* **237**, no.4 (1977) 56.

Cline D.B., Mann A.K. and Rubbia C. : The search for new families of elementary particles. *Sci. Am.* **234**, no.1 (1966) 44.

Glashow S.L. : Quarks with color and flavor. *Sci. Am.* **233**, no.4 (1975) 38.

Gasiorowicz S. and Rosner J.L. : Hadron spectra and quarks. *Am. J. Phys.* **49** (1981) 954.

Rosner J.L. : Resource letter NP-I: New particles. *Am. J. Phys.* **49** (1980) 90.

Gottfried K. : Are they the hydrogen atoms of strong-interaction physics? *CONAPP* **9** (1981) 141.

Glashow S.L. : A question of flavor. *CONAPP* **8** (1978) 21.

Lederman L.M. : New quarks and old ones, too. *CONAPP* **8** (1978) 45.

Ellis J. : Bottomology and topology. *CONAPP* **8** (1978) 21.

Quigg C. and Rosner J.L. : Scaling the Schrödinger equation (for quarkonium). CONAPP **8** 1978) 11.

Goldhaber G. : The case for charmed mesons. *CONAPP* **7** (1977) 97.

Lederman L.M. : Leptomania. *CONAPP* **7** (1977) 89.

Perl M. : The total cross section in e^+e^- annihilation and the new particles. *CONAPP* **7** (1977) 55.

Treiman S.B. : Heavy leptons. *CONAPP* **7** (1977) 35.

Lee B.W., Quigg C. and Rosner J.L. : Tests for weak decays of charmed particle candidates. *CONAPP* **7** (1977) 49.

Harari H. : Three quarks are not enough. Do we need four? Six? More? *CONAPP* **6** (1976) 123.

Barish B.G. : Evidence for new physics in high energy neutrino collisions (dimuons). *CONAPP* **6** (1976) 87.

Cline D. and Mann A.K. : A new quantum number of hadronic matter. *CONAPP* **6** (1976) 75.

Richter B. : From the psi to charm: the experiemtns of 1975 and 1976 (The 1976 Nobel Prize address). *Rev. Mod. Phys.* **49** (1977) 251.

Ting S.C.C. The discovery of the J particle: a personal recollection (The 1976 Nobel Prize address). *Rev. Mod. Phys.* **49** (1977)235.

Franzini P. and Lee-Franzini J. : Upsilon physics at CESR. *Phys. Rep.* **81** (1982) 239.

Trilling G.H. : The properties of charmed particles. *Phys. Rep.* **75** (1981) 57.

Veisenberg A.O. : Lifetime of charmed particles (a review of experimental data). *Sov. Phys. Uspekhi* **24** (1981) 733.

Azimov Ya.A. and Khoze V.A. : Present status of the τ lepton. *Sov. Phys. Uspekhi* **23** (1980) 699.

Azimov L.N., Frankfurt L.L. and Khoze V.A. : New particle in e^+e^- annihilation: heavy lepton τ. *Sov. Phys. Uspekhi* **21** (1978) 225.

Bunyatov S.A. : A search for new particles and antinuclei. *Sov. J. Part. Nucl.* **10** (1979) No.4.

Bogolyubov P.N. et al. : New particles and their possible interpretations. *Sov. J. Part. Nucl.* **7** (1976) 325.

3. The Electromagnetic Interaction

3.1. Quantum electrodynamics

Ekstrom P. and Wineland D. : The isolated electron. *Sci. Am.* **243**, no.2 (1980) 90.

Fulcher L.P., Rafelski J. and Klein A. : The decay of the vacuum. *Sci. Am.* **241**, no.6 (1979) 120.

Hänsch T.W., Schawlow A.L. and Series G.W. : The spectrum of atomic hydrogen. *Sci. Am.* **240**, no.3 (1979) 72.

Goldhaber A.S. and Nieto M.M. : The mass of the photon. *Sci. Am.* **234**, no.5 (1976) 86.

Bederson B. : Atomic physics: a renewed vitality. *Phys. Today* **34** (1981) 188.

Rich A. : Recent experimental advances in positronium research. *Rev. Mod. Phys.* **53** (1981) 127.

Calmet J. et al. : The anomalous magnetic moment of the muon. A review of the theoretical contributions. *Rev. Mod. Phys.* **49** (1977) 21.

Soffel M., Muller B. and Greiner W. : Stability and decay of the Dirac vacuum in external gauge fields. *Phys. Rep.* **85** (1982) 51.

Baier V.N., Fadin V.S., Khoze V.A. and Kuraev E.A. : Inelastic processes in high energy quantum electrodynamics. *Phys. Rep.* **78** (1981) 293.

Combley F., Farley F.J.M. and Picasso E. : The CERN muon (g-2) experiments. *Phys. Rep.* **68** (1981) 93.

Bodwin G.T. and Yennie D.R. : Hyperfine splitting in positronium and muonium. *Phys. Rep.* **43** (1978) 267.

Rafelski J., Fulcher L.P. and Kelin A. : Fermions and bosons interacting with arbitrary strong external fields. *Phys. Rep.* **38** (1978) 228.

Stroscio M.A. : Positronium: a review of the theory. *Phys. Rep.* **21** (1975) 81.

Kadyshevsky V.G. : New approach to the theory of electromagnetic interactions with a fundamental length. *Sov. J. Part. Nucl.* **11** (1980) 1.

Fomin P.I. : Quantum electrodynamics at short distances. *Sov. J. Part. Nucl.* **7** (1976) 269.

3.2. e^+e^- collisions (see also 1.1, 2.5, 5.5)

Jacob M. : Physics at LEP. *CONAPP* **9** (1980) 31.

Schopper H. : Two years of PETRA operation. *CONAPP* **10** (1981) 33.

Criegee L. and Knies G. : e^+e^- physics with the PLUTO detector. *Phys. Rep.* **83** (1982) 151.

The Mark J collaboration: Physics with high energy electron–positron colliding beams with the Mark J detector. *Phys. Rep.* **63** (1980) 337.

Perez-y-Jorba J.P. and Renard F.M. : The physics of electron-positron colliding beams. *Phys. Rep.* **31** (1977) 1.

Ioffe B.L. and Khoze V.A. : What can be expected from experiments with e^+e^- colliding beams at energy $\sim$ 100 GeV. *Sov. J. Part. Nucl.* **9** (1978) 50.

Bazhanov V.V., Pron'ko G.P. and Solov'ev L.D. : Electromagnetic effects in resonance processes in colliding beams. *Sov. J. Part. Nucl.* **8** (1977) 1.

3.3. Electromagnetic interactions of hadrons

Jacob M. and Landshoff P. : The inner structure of the proton. *Sci. Am.* **242**, no.3 (1980) 46.

Peoples J. and Lee W. : Photoproduction experiments at Fermilab. *CONAPP* **6** (1976) 83.

O'Donnel P.J. : Radiative decays of (old) mesons. *Rev. Mod. Phys.* **53** (1981) 673.

Baner T.H. et al. : The hadronic properties of the photon in high-energy interactions. *Rev. Mod. Phys.* **50** (1978) 261.

Stroynowsky R. : Lepton pair production in hadron collisions. *Phys. Rep.* **71** (1981) 1.

Francis W.R. and Kirk T.B.W. : Muon scattering at Fermilab. *Phys. Rep.* **54** (1979) 307.

Craigie N.S. : Lepton an photon production in hadron collisions. *Phys. Rep.* **47** (1978) 1.

Isayev P.P. : Interaction. *Fiz. Elem. Chastits At. Yadra* **13** (1982) 82.

Petrun'kin V.A. : Electric and magnetic polarizability of hadrons. *Sov. J. Part. Nucl.* **12** (1981) 278.

Govorkov B.B. : Investigation of electromagnetic interactions by means of electron–photon beams from proton accelerators. *Sov. J. Part. Nucl.* **11** (1980) 442.

Baranov P.S. and Fil'kov L.V. : Compton scattering on the proton at low and medium energies. *Sov. J. Part. Nucl.* **7** (1976) 42.

4. The Strong Interaction

4.1. Quantum chromodynamics (see also 11.4, 11.10)

Nambu Y. : The confinement of quarks. *Sci. Am.* **235**, no.5 (1976) 48.

Drell S.D. : The Richtmyer memorial lecture—when is a particle? *Am. J. Phys.* **46** (1978) 597.

Drell S.D. : When is a particle? *Phys. Today* **31** (1978) 23.

Gross D.J. , Pisarski R.D. and Yaffe L.G. : QCD and instantons at finite temperature. *Rev. Mod. Phys.* **53** (1981) 43.

Buras A.J. : Asymptotic freedom in deep inelastic processes in the leading order and beyond. *Rev. Mod. Phys.* **52** (1980) 199.

Altarelli G. : Partons in quantum chromodynamics. *Phys. Rep.* **81** (1982) 1.

Bander M. : Theories of quark confinement. *Phys. Rep.* **75** (1981) 205.

Mueller A.N. : Perturbative QCD at high energies. *Phys. Rep.* **73** (1981) 237.

Reya E. : Perturbative quantum chromodynamics. *Phys. Rep.* **69** (1981) 195.

Dokshitzer Yu.L. , Dyakonov D.I. and Troyan S.I. : Hard processes in quantum chromodynamics. *Phys. Rep.* **58** (1980) 269.

Marciano W. and Pagels H. : Quantum chromodynamics. *Phys. Rep.* **36** (1978) 137.

Vainshtein A.I. , Zakharov V.I. , Novikov V.A. and Shifman M.A. : ABC of instantons. *Sov. Phys. Uspekhi* **24** (1982) 195.

Vainshtein A.I. , Zakharov V.I. , Novikov V.A. and Shifman M.A. : Quantum chromodynamics and hadron mass scales. *Fiz. Elem. Chastits At. Yadra* **13** (1982) 542 (in Russian).

4.2. Gluons

Donoghue J.F. : Glueballs. *CONAPP* **10** (1982) 277.
Ellis J. : Gluons. *CONAPP* **9** (1981) 153.
Duinker P. and Luckey D. : In search of gluons. *CONAPP* **9** (1980) 123.
Michael C. : Perturbative tests of quantum chromodynamics. *CONAPP* **8** (1978) 89.

Azimov I.M. , Dokshitzer Yu.L. and Khoze V.A. : Gluons. *Sov. Phys. Uspekhi* **23** (1980) 732.
Dremin I.M. : Gluon jets. *Sov. Phys. Uspekhi* **23** (1980) 515.

4.3. Quarkonia (see also 2.5, 3.2)

Bloom H. and Martin A. : Exact results on potential models for quarkonium systems. *Phys. Rep.* **60** (1980) 341.
Quigg C. and Rosner J.L. : Quantum mechanics with applications to quarkonium. *Phys. Rep.* **56** (1979) 167.
Novikov V.A. et al. : Charmonium and gluons. *Phys. Rep.* **41** (1978) 1.

Filippov A.T. : Spectroscopy of light measons. *Sov. Phys. Uspekhi* **25** (1982) 371.
Vainshtein A.I. , Voloshin M.B. , Zakharov V.I. , Novikov V.A. , Okun L.B. and Shifman M.A. : Charmonium and quantum chromodynamics. *Sov. Phys. Uspekhi* **20** (1977) 796.

4.4. Bag model

Miller G.A. , Theberger S. and Thomas A.W. : Pionic corrections to the MIT bag model. *CONAPP* **10** (1981) 101.

Lee T.D. : Nontopological solitons and applications to hadrons. *CONAPP* **7** (1978) 165.

Jaffe R.L. and Johnson K. : A practical model of quark confinement. *CONAPP* **7** (1977) 107.

Kuti J. and Hasenfratz P. : The quark mode bag model. *(Phys. Rep.* **40** (1978) 75.

4.5. Free quarks

Fairbank W.M. Jr. and Franklin A. : Did Millikan observe fractional charges on oil drops? *Am. J. Phys.* **50** (1982) 394.

Pati J.C. and Salam A. : The unconfined unstable quark. Part II. *CONAPP* **7** (1976) 1.

Pati J.C. and Salam A. : The unconfined unstable quark. Part I. *CONAPP* **6** (1976) 183.

Okun L.B. and Zeldovich Ya. B. : Realistic quark models and astrophysics. *CONAPP* **6** (1976) 69.

Jones L.W. : A review of quark search experiments. *Rev. Mod. Phys.* **49** (1977) 717.

Marinelli M. and Morpurgo G. : Searches of fractionally charged particles in matter with the magnetic levitation technique. *Phys. Rep.* **85** (1982) 161.

Arbuzov B.A. and Tikhonin F.F. : Weak interaction in quark models with unconfined color. *Sov. J. Part. Nucl.* **11** (1980) 423.

Govorkov A.B. : Color degrees of freedom in hadron physics. *Sov. J. Part. Nucl.* **8** (1977) 431.

4.6. Baryonium. Dibaryons. Hadronic atoms

Provh B. : The barionium—exotic nuclear state or new boson resonance? *CONAPP* **8** (1978) 69.

Badalyan A.M. et al. : Resonance in coupled channels in nuclear and particle physics. *Phys. Rep.* **82** (1982) 31.

Montanet L., Rossi G.C. and Veneziano G. : Baryonium physics. *Phys. Rep.* **63** (1980) 149.

Shaprio I.S. : The physics of nucleaon– antinucleon systems. *Phys. Rep.* **35** (1978) 129.

Simonov Yu.A. : Theoretical interpretation of dibaryon experimental data. *Sov. Phys. Uspekhi* **25** (1982) 99.

Makarov M.M : Dibaryon resonances. *Sov. Phys. Uspekhi* **25** (1982) 185.

Betty S.J. : Exotic atoms. *Fiz. Elem. Chastits At. Yadra* **13** (1982) 164 (in Russian).

4.7. Processes involving a small number of particles

Krish A.D. : The spin of the proton. *Sci. Am.* **240**, no.5 (1979) 58.

Mukhin N.N. and Patarakin O.O. : The pion–pion interaction (review of experimental data). *Sov. Phys. Uspekhi* **24** (1981) 161.

Bel'kov A.A. and Bunyatov S.A. : Study of pion–pion interaction at low energies in $\pi N \to \pi\pi N$ reactions close to the threshold. *Fiz. Elem. Chastits At. Yadra* **13** (1982) 5 (in Russian).

Volkov M.K. : Decays of the fundamental meson octet in quantum chiral theory. *Sov. J. Part. Nucl.* **10** (1979) No.4.

Nikitin B.A. : Experimental study of binary reactions. *Sov. J. Part. Nucl.* **10** (1979) No.4.

Zotov N.P. and Tsarev V.A. : Diffraction dissociation and the Drell–Hiida–Deck model. *Sov. J. Part. Nucl.* **9** (1978) 266.

Lapidus L.I. : Polarization phenomena in hadron collisions at low momentum transfers. *Sov. J. Part. Nucl.* **9** (1978) 35.

Achasov N.N. and Shestakov P.N. : Effects of ρ°-w mixing and dynamics of vector meson production. *Sov. J. Part. Nucl.* **9** (1978) 19.

Gaĭsak M.I. and Lend'el V.I. : Description of low energy πN scattering in nonlinear $SU(2) \times SU(2)$ chiral dynamics. *Sov. J. Part. Nucl.* **8** (1977) 452.

Mukhin S.V. and Tsarev V.A. : Diffraction excitation of protons on protons and deuterons at high energies and small momentum transfers. *Sov. J. Part. Nucl.* **8** (1977) 403.

Savin I.A. : $K_L^\circ \to K_S^\circ$ transmission regeneration in hydrogen. *Sov. J. Part. Nucl.* **8** (1977) 11.

Ponomarev L.N. : Description of exclusive processes in the Reggeized one-pion-exchange model. *Sov. J. Part. Nucl.* **7** (1976) 70.

4.8. Multiple production

Jacob M. : Ins physics, knowledge and problems. *CONAPP* **6** (1979) 133.

Jones L.M. : Pion exchange at high energies. *Rev. Mod. Phys.* **52** (1980) 545.

Abarbanel H.D.I. : Diffraction scattering of hadrons: the theoretical outlook. *Rev. Mod. Phys.* **48** (1976) 435.

Kane S.L. and Seidl A. : An interpretation of two-body hadron reactions. *Rev. Mod. Phys.* **48** (1976) 309.

Fischer J. : General laws of hadron scattering at very high energies. *Phys. Rep.* **76** (1981) 157.

Ganguli S.N. and Roy D.P. : Regge phenomenology of inclusive reactions. *Phys. Rep.* **67** (1980) 201.

Giacomelli G. and Jacob M. : Physics at the CERN-ISR. *Phys. Rep.* **55** (1979) 1.

Kaidalov A.B. : Diffractive production mechanisms. *Phys. Rep.* **50** (1979) 157.

Irving A.C. and Worden R.P. : Regge phenomenology. *Phys. Rep.* **34** (1977) 117.

Feinberg E.L. : Hadron clusters and half-dressed particles in quantum field theory. *Sov. Phys. Uspekhi* **23** (1980) 629.

Grishin V.G. : Inclusive processes in high-energy hadron interactions. *Sov. Phys. Uspekhi* **22** (1979) 1.

Dremin I.M. and Quigg C. : Clusters in hadron multiple production processes. *Sov. Phys. Uspekhi* **21** (1978) 265.

Kladnitskaya E.N. : Production of neutral strange particles in π^-p and pp high-energy interactions. *Fiz. Elem. Chastits At. Yadra* **13** (1982) 669 (in Russian).

Budagov Yu. A. et al. : Multiple production of neutral particles in pion–proton and proton–proton interactions. *Sov. J. Part. Nucl.* **11** (1980) 273.

Dremin I.M. and Feinberg E.L. : Clusters in multiparticle production. *Sov. J. Part. Nucl.* **10** (1979) 394.

Grishin V.G. : High-energy multiple-particle reactions. *Sov. J. Part. Nucl.* **10** (1979) No.4.

Tyapkin A.A. : Statistical theory of multiple production of hadrons. *Sov. J. Part. Nucl.* **8** (1977) 222.

Grishin V.G. : Multiple production of particles in hadron–hadron collisions at high energies. *Sov. J. Part. Nucl.* **7** (1976) 233.

4.9. Large momentum transfer. Jets

Jacob M. : Hadronic jets. *CONAPP* **8** (1978) 1.

Ellis S.D. and Stroynowski R. : Large p_T physics: data and the constituent models. *Rev. Mod. Phys.* **49** (1977) 753.

Jacob M. and Landshoff P. : Large transverse momentum and jet studies. *Phys. Rep.* **48** (1978) 285.

Goloskokov S.V. , Kudinov A.V. and Kuleshov S.P. : Preasymptotic effects in high-energy large-angle elastic scattering of hadrons. *Sov. J. Part. Nucl.* **12** (1981) 248.

Zotov N.P. , Rusakov S.V. and Tsarev V.A. : Elastic proton–proton scattering at high energies and large momentum transfers. *Sov. J. Part. Nucl.* **11** (1980) 462.

Geyer B. , Robaschik D. and Wieczorek E. : Field-theoretical description of deep inelastic scattering. *Sov. J. Part. Nucl.* **11** (1980) 52.

Ranft G. and Ranft J. : Production of particles with large transverse momenta and models of hard collisions. *Sov. J. Part. Nucl.* **10** (1979) 35.

Goloskokov S.V. et al. : Asymptotic power-law scaling of large-angle hadron-hadron scattering. *Sov. J. Part. Nucl.* **8** (1977) 395.

Kvinikhidze A.N. et al. : Inclusive processes with large transverse momenta in the composite particles approach. *Sov. J. Part. Nucl.* **8** (1977) 196.

Bilen'kii S.M. : Deep inelastic neutrino processes. *Sov. J. Part. Nucl.* **8** (1977) 30.

4.10. High-energy interactions in cosmic rays

Gaisser T.K. et al. : Cosmic ray showers and particle physics at energies 10^{15}–10^{18} eV. *Rev. Mod. Phys.* **50** (1978) 859.

Lattes C.M.G. , Fujimoto Y. and Hasegawa S. : Hadronic interactions of high energy cosmic rays observed by emulsion chambers. *Phys. Rep.*, **65** (1980) 151.

Nikol'skii S.I. : Hadronic interactions in cosmic rays at energies above the accelerator range. *Sov. Phys. Uspekhi* **24** (1981) 925.

4.11. High energy interactions of hadrons with nuclei

Schiffer J.P. : Pion (or Δ resonance) absorption in nuclear medium—a synthesis of pion- and photo-nuclear reactions. *CONAPP* **10** (1981) 343.

Kisslinger L.S. : Nuclear and particle physics at energies up to 31 GeV. *CONAPP* **10** (1981) 187.

Bergström L. and Fredriksson S. : The deuteron in high-energy physics. *Rev. Mod. Phys.* **52** (1980) 675.

Igo G.J. : Some recent intermediate- and high-energy proton-nucleus research. *Rev. Mod. Phys.* **50** (1978) 523.

Frankfurt L.L. and Strikman M.I. : High-energy phenomena, short-range nuclear structure and QCD. *Phys. Rep.* **76** (1981) 215.

Nikolaev N.N. : Quarks in high-energy interactions of hadrons, photons, and leptons with nuclei. *Sov. Phys. Uspekhi* **24** (1981) 531.

Efremov A.V. : Quark–parton pattern of cumulative production. *Fiz. Elem. Chastits At. Yadra* **13** (1982) 613 (in Russian).

Amelin N.S. , Glagolev V.V. and Lykasov G.I. : Characteristic features of hadron interaction with light nuclei at medium energies. *Fiz. Elem. Chastits At. Yadra* **13** (1982) 130 (in Russian).

Shabel'skii Yu.M. : Multiple production in hadron– nucleus collisions at high energies. *Sov. J. Part. Nucl.* **12** (1981) 430.

Nikolaev N.N. : Interactions of high-energy hadrons, photons, and leptons with nuclei. *Sov. J. Part. Nucl.* **12** (1981) 63.

Kalinkin B.N. and Shmonin V.L. : Space–time description of multiparticle production in nuclear matter and the structure of hadrons. *Sov. J. Part. Nucl.* 11 (1980) 248.

Strikman M.I. and Frankfurt L.L. : Scattering of high-energy particles as a probe of few-nucleon correlations in the deuteron and in nuclei. *Sov. J. Part. Nucl.* **11** (1980) 221.

Stavinskii V.S. : Limiting fragmentation of nuclei—the cumulative effect (experiment). *Sov. J. Part. Nucl.* **10** (1979) 373.

Lukyanov V.K. and Titov A.I. : Large-momentum-transfer nuclear reactions and nuclear flucton hypothesis. *Sov. J. Part. Nucl.* **10** (1979) No.4.

Gulamov K.G. , Chernov G.M. and Gulamov U.G. : Experimental data on multiparticle production on nuclei. *Sov. J. Part. Nucl.* **9** (1978) 226.

Baldin A.M. : Physics of relativistic nuclei. *Sov. J. Part. Nucl.* **8** (1977) 175.

Tarasov A.V. : Coherent and incoherent production of particles on nuclei in the theory of multiple scattering. *Sov. J. Part. Nucl.* **7** (1976) 306.

4.12. Nuclear matter

Rho M. and Brown G.E. : The role of chiral invariance in nuclei. *CONAPP* **10** (1981) 201.

Johnson M.B. and Bethe H.A. : Nuclear matter distributions with pions. *CONAPP* **8** (1978) 75.

Migdal A.B. : Pion fields in nuclear matter. *Rev. Mod. Phys.* **50** (1978) 107.

Oset E. , Tobi H. and Weise W. : Pionic models of excitation in nuclei. *Phys. Rep.* **8** (1982) 281.

Shuryak E.V. : Quantum chromodynamics and the theory of superdense matter. *Phys. Rep.* **61** (1980) 70.

Morley P.D. and Kisslinger M.B. : Relativistic many-body theory, quantum chromodynamics and neutron stars/supernova. *(Phys. Rep.* **51** (1979) 63.

4.13. Semi-phenomenological approaches to strong interactions

Capra F. : Quark physics without quarks. A review of recent development in S-matrix theory. *Am. J. Phys.* **47** (1979) 11.

Cocconi G. : Dimensional considerations about elementary particles. *CONAPP* **7** (1978) 177.

Chew G.F. and Rosenzweig C. : Dual topological unitarization: an ordered approach to hadron theory. *Phys. Rep.* **41** (1978) 263.

Amirkhanov I.V. , Grusha G.V. and Mir-Kasimov R.M. : Quasipotential equation in terms of rapidities and applications to relativistic bound-state and scattering problems. *Sov. J. Part. Nucl.* **12** (1981) 262.

Kirzhnits D.A. , Kryuchkov G. Yu. and Takibayev N.Zh. : A new approach to quantum theory and its applications to low- and high-energy nuclear physics. *(Sov. J. Part. Nucl.* **10** (1979) No.4.

Pišut J. : Statistical approach to analytic extrapolations in strong interactions. *Sov. J. Part. Nucl.* **9** (1978) 246.

Skachkov N.B. and Solobtsov I.L. : Relativistic three-dimensional description of the interaction of two fermions. *Sov. J. Part. Nucl.* **9** (1978) 1.

Savrin V.I. , Tyurin N.E. and Khrustalev O.A. : U-matrix method in the theory of strong interactions. *Sov. J. Part. Nucl.* **7** (1976) 9.

Ginzburg V.L. and Man'ko V.I. : Relativistic wave equations with internal degrees of freedom and partons. *Sov. J. Part. Nucl.* **7** (1966) 1.

5. The Weak Interaction

5.1. Weak processes (see also 2.3, 2.5)

Holstein B.R. : Resource letter W-I: Weak interactions. *Am. J. Phys.* **45** (1977) 1033.

Telegdi V.L. : Can we still strongly believe in weak magnetism? *CONAPP* **8** (1979) 171.

Devlin T.J. and Dickey J.O. : Weak hadronic decays: K $\to$ 2π and K $\to$ 3π. *Rev. Mod. Phys.* **51** (1979) 237.

Fritzsch H. and Minkowski P. : Flavordynamics of quarks and leptons. *Phys. Rep.* **7** (1981) 67.

Barish B.C. : Experimental aspects of high energy neutrino physics. *Phys. Rep.* **39** (1978) 279.

Musset P. and Vialle J.P. : Neutrino physics with Gargamelle. *Phys. Rep.* **39** (1978) 1.

Ryder L.H. : What can low-energy nuclear physics tell us about elementary particles and quarks? *Phys. Rep.* **34** (1977) 55.

Sushkov O.I. and Flambaum V.V. : Parity breaking in the interaction of neutrons with heavy nuclei. *Sov. Phys. Uspekhi* **25** (1982) 1.

Kopeliovich V.B. : New results on the violation of parity in proton-proton and nucleon-nucleus interactions. *Sov. Phys. Uspekhi* **24** (1981) 717.

Danilyan G.V. : Parity violation in nuclear fission. *Sov. Phys. Uspekhi* **23** (1980) 323.

Ermolov P.F. and Mukhin A.I. : Neutrino experiments at high energies. *Sov. Phys. Uspekhi* **21** (1978) 185.

Bardin D.Yu. and Ivanov E.A. : Weak-electromagnetic decays $\pi(K) \to l\nu\gamma$ and $\pi(K) \to l\nu l'^{+}l'^{-}$. *Sov. J. Part. Nucl.* **7** (1976) 286.

5.2. CP-violation (see also 2.4)

Cronin J.W. : CP symmetry violation—the search for its origin (The 1980 Nobel Prize address). *Rev. Mod. Phys.* **53** (1981) 373.

Fitch V.L. : The discovery of charge-conjugation parity asymmetry, (The 1980 Nobel Prize address). *Rev. Mod. Phys.* **53** (1981) 367.

Krasnikov N.V. , Matveev V.A. and Tavkhelidze A.N. : The problem of CP invariance in quantum chromodynamics. *Sov. J. Part. Nucl.* **12** (1981) 38.

5.3. Neutral currents

Cline D., Mann A.K. and Rubbia C. : The detection of neutral weak currents. *Sci. Am.* **231**, no.6 (1974) 108.

Batlay C. : The status of weak currents. *CONAPP* **8** (1979) 157.
Feinberg G. : Why all the fuss about bismuth? *CONAPP* **8** (1979) 143.
Paschos E.A. : Neutral currents. *CONAPP* **7** (1977) 153.
Henley E.M. : Do the weak neutral currents cause parity nonconserving eN and μN forces? *CONAPP* **7** (1977) 79.

Kim J.E. et al. : Theoretical and experimental review of the weak neutral current: a determination of its structure and limits on deviations from the minimal $SU(2)_L \times U(I)$ electroweak theory. *Rev. Mod. Phys.* **53** (1981) 211.

Donnelly T.W. and Peccei R.D. : Neutral current effects in nuclei. *Phys. Rep.* **50** (1979) 1.

Barkov L.M. , Zolotorev M.S. and Khriplovich I.B. : Observation of parity nonconservation in atoms. *Sov. Phys. Uspekhi* **23** (1980) 713.

Shekhter V.M. : Weak interaction involving neutral currents. *Sov. Phys. Uspekhi* **19** (1976) 645.

Moskalev A.N., Rynding R.M. and Khriplovich I.B. : Possible lines of research into weak-interaction effects in atomic physics. *Sov. Phys. Uspekhi* **19** (1976) 220.

Alekseev V.A. , Zel'dovich Ya.B. and Sobelman I.I. : Parity nonconservation effects in atoms. *Sov. Phys. Uspekhi* **19** (1976) 207.

5.4. Double beta-decay

Bryman D. and Picciotto C. : Double beta decay. *Rev. Mod. Phys.* **50** (1978) 11.

Zdesenko Yu.G. : Double β-decay and lepton charge conservation. *Sov. J. Part. Nucl.* **11** (1980) No.6

5.5. Electroweak theory (see also 1.1, 3.2, 11.4)

Cline D.B. , Rubbia C. and van der Meer S. : The search for intermediate vector bosons. *Sci. Am.* **246**, no.3 (1982) 48.

Paschos E.A. and Wang L.L. : The quest for W's. *CONAPP* **6** (1976) 115.
Glashow S.L. : Towards a unified theory: threads into a tapestry (The 1979 Nobel Prize address). *Rev. Mod. Phys.* **52** (1980) 539.
Salam A. : Gauge unification of fundamental forces (The 1979 Nobel Prize address). *Rev. Mod. Phys.* **52** (1980) 525.
Weingberg S. : Conceptual foundations of the unified theory of weak and electromagnetic interactions (The 1979 Nobel Prize address). *Rev. Mod. Phys.* **52** (1980) 515.
Sirlin A. : Current algebra formulation in gauge theories and the universality of the weak interactions. *Rev. Mod. Phys.* **50** (1978) 573.
Quigg C. : Production and detection of invermediate vector bosons and heavy leptons in pp and $\bar{p}p$ collisions. *Rev. Mod. Phys.* **49** (1977) 297.

Volkov G.G. , Liparteliani A.G. and Nikitin Yu.P : Gauge schemes of weak and electromagnetic interactions. *Sov. J. Part. Nucl.* **10** (1979) 71.

6. Scalar Bosons and Technicolor (see also 11.4)
Gaillard M.K. : The Higgs particle. *CONAPP* **8** (1978) 31.

Farri E. and Susskind L. : Technicolor. *Phys. Rep.* **74** (1981) 277.

Vainshtein A.I. , Zakharov V.I. and Shifman M.A. : Higgs particles. *Sov. Phys. Uspekhi* **23** (1980) 429.

7. Grand Unification (see also 10.3, 11.4)
Carrigan R.A. Jr. and Trower W.P. : Superheavy magnetic monopoles. *Sci. Am.* **246**, no.4 (1982) 106.
Weinberg S. : The decay of the proton. *(Sci. Am.* **244**, no.6 (1981) 52.
Georgi H. : Unified theory of elementary particles and forces. *Sci. Am.* **244**, no.4 (1981) 40.

Bergmann P. : Unitary field theories. *Phys. Today* **32**, no.3 (1979) 44.

Weinberg S. : The future of unified gauge theories. *Phys. Today* **30**, no.4 (1977) 42.

Goldhaber M. and Sulak L.R. : An overview of current experiments in search of proton decay. *CONAPP* **10** (1981) 215.

Gell-Mann M , Ramond R. and Slansky R. : Color embeddings, charge assignments and proton stability in unified gauge theories. *Rev. Mod. Phys.* **50** (1978) 721.

Rossi P. : Exact results in the theory of non-abelian magnetic monopoles. *Phys. Rep.* **86** (1982) 317.
Slansky R. : Group theory for unified model building. *Phys. Rep.* **79** (1981) 1.
Langacker P. : Grand unified theories and proton decay. *Phys. Rep.* **72** (1981) 185.
Harari H. : Quarks and leptons. *Phys. Rep.* **42** (1978) 235.

Matinyan S.G. : Toward the unification of weak, electromagnetic, and strong interactions. *Sov. Phys. Uspekhi* **23** (1980) 11.
Arbuzov B.A. and Logunov A.A. : Structure of elementary particles and relationships between the different forces of nature. *Sov. Phys. Uspekhi* **20** (1977) 956.

8. Supersymmetry and Supergravity

Freedman D.Z. and van Nieuwenhuizen P. : Supergravity and unification of the laws of physics. *Sci. Am.* **238**, no.2 (1978) 126.

Van Nieuwenhuizen P. : Supergravity. *Phys. Rep.* **68** (1981) 189.
Fayet P. and Ferrara S. : Supersymmetry. *Phys. Rep.* **32** (1977) 249.

Slavnov A.A. : Supersymmetric gauge theories and their possible applications to the weak and electromagnetic interactions. *Sov. Phys. Uspekhi* **21** (1978) 240.
Ogievetskii V.I. and Mezincescu L. : Boson-fermion symmetries and superfields. *Sov. Phys. Uspekhi* **18** (1975) 960.

9. Gravitation (see also 8, 10, 11)

9.1. General relativity

Weisberg J. , Taylor J.H. and Fowler L.A. : Gravitational waves from an orbiting pulsar. *Sci. Am.* **245**, no.4 (1981) 66.

Chaffe F.H. Jr. : The discovery of gravitational lens. *Sci. Am.* **243**, no.5 (1980) 60.
Callahan J.J. : The curvature of space in a finite universe. *Sci. Am.* **235**, no.2 (1976) 90.
Van Flandern T.C. : Is gravity getting weaker? *Sci. Am.* **234**, no.2 (1976) 44.

Worden P.W. Jr. and Everitt C.W.F. : Resource letter GI-1: Gravity and Inertia. *Am. J. Phys.* **50** (1982) 494.
Higbie J. : Gravitational lens. *Am. J. Phys.* **49** (1981) 652.
di Sessa A.A. : An elementary formalism for general relativity. *Am. J. Phys.* **49** (1981) 401.

Wesson P.S. : Does gravity change with time? *Phys. Today* **33** (1980) 32.

Thorne K.S. : Multipole expansions of gravitational radiation. *Rev. Mod. Phys.* **52** (1980) 299.
Thorne K.S. : Gravitational-wave research: current status and future prospects. *Rev. Mod. Phys.* **52** (1980) 285.
Hehl F.W. et al. : General relativity with spin and torsions: foundations and prospects. *Rev. Mod. Phys.* **48** (1976) 393.

Barrow J.D. : Chaotic behavior in general relativity. *Phys. Rep.* **85** (1982) 1.
Carmeli M., Charach Ch. and Malin S. : Survey of cosmological models with gravitational, scalar and electromagnetic waves. *Phys. Rep.* **76** (1981) 79.
Marsden J.E. and Tipler F.J. : Maximal hypersurfaces and foliations of constant mean curvature in general relativity. *Phys. Rep.* **66** (1980) 109.
Kandrup H.E. : Stochastic gravitational fluctuations in a self-consistent mean field theory. *Phys. Rep.* **63** (1980) 1.
Sivaram C. and Sinha K.P. : Strong spin-two interaction and general relativity. *Phys. Rep.* **51** (1979) 111.
Braginsky V.B. and Rudenko V.N. : Gravitational waves and the detection fo gravitational radiation. *Phys. Rep.* **46** (1978) 165.
Papini G. and Valuri S.R. : Gravitons in Minkowski space–time. Interactions and results of astrophysical interest. *Phys. Rep.* **33** (1977) 51.

Faddeev L.D. : The energy problem in Einstein's theory of gravitation. *Sov. Phys. Uspekhi* **25** (1982) 435.

Mukhanov V.F. : The double quasar QSO-0957+561 A,B: a gravitational lens? *Sov. Phys. Uspekhi* **24** (1981) 331.

Hawking S.W. and Israel W. : Introductory survey. In: *An Einstein Centenary Survey*, edited by Hawking S.W. and Israel W. Cambridge, Cambridge University Press (1979) p. 1.

Rudenko V.N. : Relativistic experiments in gravitational fields. *Sov. Phys. Uspekhi* **21** (1978) 893.

Konopleva N.P. : Gravitational experiments in space. *Sov. Phys. Uspekhi* **20** (1977) 537.

Denisov V.I. and Logunov A.A. : A new theory of space–time and gravitation. *Fiz. Elem. Chastits At. Yadra* **13** (1982) 157.

Denisov V.I. , Logunov A.A. and Mestvirishvili M.A. : A field theory of gravitation and new notions of space and time. *Sov. J. Part. Nucl.* **12** (1981) 1.

Alekseev G.A. and Khlebnikov V.I. : The Newman–Penrose formalism and its application in the general theory of relativity. *Sov. J. Part. Nucl.* **9** (1978) 421.

Volovich I.V., Zagrebnov V.A. and Frolov V.P. : Quantum Field theory in asymptotically flat spacetime. *Sov. J. Part. Nucl.* **9** (1978) 63.

9.2. Black holes

Hawking S.W. : The quantum mechanics of black holes. *Sci. Am.* **236**, no.1 (1977) 34.

Doughty N.A. : Surface properties of Kerr–Newman black holes. *Am. J. Phys.* **49** (1981) 720.

Doughty N.A. : Acceleration of a static observer near the event horizon of a static isolated black hole. *Am. J. Phys.* **49** (1981) 412.

Detweiler S. : Resource Letter BH-I: Black holes. *Am. J. Phys.* **49**. (1981) 394.

Brehme R.W. : Inside the black hole. *Am. J. Phys.* **45** (1977) 423.

Bekenstein J.D. : Black-hole thermodynamics. *Phys. Today* **33** no.1 (1980) 24.

Frolov V.P. : Quantum theory of gravitation. (Survey of the 2nd International Seminar on Quantum Gravity, Moscow, 13–15 October 1981). *Usp. Fiz. Nauk* **138** (1982) 151.

10. Astrophysics and Cosmology

10.1. Astrophysics

de Boer K.S and Savage B.D. : The coronas of Galaxies. *Sci. Am.* **247**, no.2 (1982) 52.

Lada C.J. : Energetic outflows from young stars. *Sci. Am.* **247**, no.1 (1982) 74.

Blandford R.D., Begelman M.C. and Rees M.J. : Cosmic jets. *Sci. Am.* **246**, no.5 (1982) 84.

Leventhal M. and MacCallum C.J. : Gamma-Ray-Line astronomy. *Sci. Am.* **243**, no.1 (1980) 50.

Giacconi R. : The Einstein X-ray observatory. *Sci. Am.* **242**, no.2 (1980) 70.

Herbst W. and Assousa G.E. : Supernovas and star formation. *Sci. Am.* **241**, no.2 (1979) 122.

Veverka J. : Phobos and Deimos. *Sci. Am.* **236**, no.2 (1977) 30.

Kirsher R.P. : Supernovas in other galaxies. *Sci. Am.* **235**, no.6 (1976) 88.

Strong I.B. and Klebsadel R.W. : Cosmic gamma-ray bursts. *Sci. Am.* **235**, no.4 (1976) 66.

Stephenson F.R. and Clark D.H. : Historical supernovas. *Sci. Am.* **234**, no.6 (1976) 100.

Cruikshank D.R. and Morrison D. : The galilean satellites of Jupiter. *Sci. Am.* **234**, no.5 (1976) 108.

Nauenberg M. and Weisskopf V.F. : Why does the sun shine? *Am. J. Phys.* **46** (1978) 23.

Giacconi R. : The Richtmyer Memorial Lecture: Progress in X-ray astronomy. *Am. J. Phys.* **44** (1976) 121.

Field G.B. : Astronomy and astrophysics for the 1980s. *Phys. Today* **35**, no.4 (1982) 46.

Harwit M. : Physicists and astronomy—will you join the dance? *Phys. Today* **34**, no.11 (1981) 172.

Lingenfelter R.E. and Ramaty R. : Gamma-ray lines: a new window to the universe. *Phys. Today* **31**, no.3 (1978) 40.

Woosley S.E., Axelrod T.S. and Weaver T.A. : Gamma-ray astronomy. *CONAPP* **9** (1981) 186.

Trinble V. : Supernovae. Part I: the events. *Rev. Mod. Phys.* **54** (1982) 1183.

Verter F. : Cosmic gamma-ray bursts. *Phys. Rep.* **81** (1982) 293.

Grigoryan L.Sh. and Saakyan G.S. : Pionization and its astrophysical aspects. *Sov. J. Part. Nucl.* **10** (1979) 428.

Kovalsky M. : Nuclear reactions in Ap stars. *Sov. J. Part. Nucl.* **8** (1977) 464.

Muradyan R.M. : Physical and astrophysical constants and their dimensional and dimensionless combinations. *Sov. J. Part. Nucl.* **8** (1977) 73.

10.2. Primary cosmic rays

Linsley J. : The highest energy cosmic rays. *Sci. Am.* **239**, no.1 (1978) 48.

Ginzburg V.L. and Ptuskin V.S. : On the origin of cosmic rays: some problems in high-energy astrophysics. *Rev. Mod. Phys.* **48** (1976) 161.

Kuzhevsky B.M. : Gamma astronomy of the Sun and study of solar cosmic rays. *Sov. Phys. Uspekhi* **25** (1982) 392.

10.3. Cosmology (see also 2.3, 7, 9)

Osmer P.S. : Quasars as probes of the distant and early universe. *Sci. Am.* **242**, no.2 (1982) 110.

Wilczek F. : The cosmic asymmetry between matter and antimatter. *Sci. Am.* **243**, no.6 (1980) 60.

Meier D.L. and Sunyaev R.A. : Primeval galaxies. *Sci. Am.* **241**, no.5 (1979) 106.

Muller R.A. : The cosmic background radiation and the new aether drift. *Sci. Am.* **238**, no.5 (1978) 68.

Groth E.J., Peebles P.J.E., Seldner M. and Seneira R.M. : Clustering of galaxies. *Sci. Am.* **237**, no.5 (1977) 76.

Gott J.R. III and Gunn J.E. : Will the universe expand forever? *Sci. Am.* **234**, no.3 (1976) 62.

Rosen J. : Extended Mach principle. *Am. J. Phys.* **49** (1981) 259.

Turner M.S. and Schramm D.N. : Cosmology and elementary particle physics. *Phys. Today* **32**, no.9 (1979) 42.

Tinsby B.M. : The cosmological constant and cosmological change. *Phys. Today* **30**, no.6 (1977) 32.

Wilczek F. : Coming attractions of SUMs and cosmology. *CONAPP* **10** (1981) 175.

Dolgov A.D. and Zeldovich Ya.B. : Cosmology and elementary particles. *Rev. Mod. Phys.* **53** (1981) 1.

Penzias A.A. : The origin of the elements (The 1978 Nobel Prize address). *Rev. Mod. Phys.* **51** (1979) 425.

Wilson R.W. : The cosmic microwave background radiation (The 1978 Nobel Prize address). *Rev. Mod. Phys.* **51** (1979) 433.

Dyson F.J. : Time without end: physics and biology in an open universe. *Rev. Mod. Phys.* **51** (1979) 447.

Fall S.M. : Galaxy correlations and cosmology. *Rev. Mod. Phys.* **51** (1979) 21.

Jones B.J.T. : The origin of galaxies: A review of recent theoretical developments and their confrontation with observation. *Rev. Mod. Phys.* **48** (1976) 107.

Zel'dovich Ya.B. and Khlopov M.Yu. : The neutrino mass in elementary-particle physics and in big-bang cosmology. *Sov. Phys. Uspekhi* **24** (1981) 755.

Zel'dovich Ya.B. : Gravitation, charges, cosmology, and coherence. *Sov. Phys. Uspekhi* **20** (1977) 945.

Einasto J.E. : The structure of galactic systems. *Sov. Phys. Uspekhi* **19** (1976) 955.

11. Quantum Mechanics and Field Theory

11.1. Quantum mechanics

Hughes R.I.G. : Quantum logic. *Sci. Am.* **245**, no.4 (1981) 146.

d'Espagnat B. : The quantum theory and reality. *Sci. Am.* **241**, no.5 (1979) 128.

Dicke R.H. : Interaction-free quantum measurements: a paradox? *Am. J. Phys.* **49** (1981) 925.

Flores J. et al. : Decay of a compound particle and the Einstein--Podolsky–Rosen argument. *Am. J. Phys.* **49** (1981) 59.

Newton R.G. : Probability interpretation of quantum mechanics. *Am. J. Phys.* **48** (1980) 1029.

Macomber J.D. : Quantum transitions without quantum jumps. *Am. J. Phys.* **45** (1977) 522.

Caves C.M. et al. : The measurement of a weak classical force coupled to a quantum-mechanical oscillator. I. Issues of principle. *Rev. Mod. Phys.* **52** (1980) 341.

Abdel-Raouf M.A. : On the variational methods for bound-state and scattering problems. *Phys. Rep.* **84** (1982) 163.

Peshkin M. : The Aharonov–Bohm effect: why it cannot be eliminated from quantum mechanics. *Phys. Rep.* **80** (1982) 375.

Dekker H. : Classical and quantum mechanics of the damped harmonic oscillator. *Phys. Rep.* **80** (1982) 1.

Cantrell C.D. and Scully M.O. : The EPR paradox revisted. *Phys. Rep.* **43** (1978) 499.

Scully M.O. , Shea R. and McCullen J.D. : State reduction in quantum mechanics: a calculational example. *Phys. Rep.* **43** (1978) 485.

Krivchenkov V.D. : Generalized coordinates in quantum mechanics. *Sov. Phys. Uspekhi* **24** (1981) 860.

Vorontsov Yu.I. : The uncertainty relation between energy and time of measurement. *Sov. Phys. Uspekhi* **24** (1981) 150.

Fushchich V.I. and Nikitin A.G. : Nonrelativistic equations of motion for particles with arbitrary spin. *Sov. J. Part. Nucl.* **12** (1981) 465.

Filippov A.T. : Singular potentials in nonrelativistic quantum theory. *Sov. J. Part. Nucl.* **10** (1979) 193.

Shirokov Yu.M. : Quantum and classical mechanics in the phase space representation. *Sov. J. Part. Nucl.* **10** (1979) 1.

Fushchich V.I. and Nikitin A.G. : Poincaré-invariant equations of motion for particles of arbitrary spin. *Sov. J. Part. Nucl.* **9** (1978) 205.

11.2. Path integral

De Witt-Morette C., Maheshwari A. and Nelson B. : Path integration in non-relativistic quantum mechanics. *Phys. Rep.* **50** (1979) 255.

Marinov M.S. : Path integrals in quantum theory: and outlook of basic concepts. *Phys. Rep.* **60** (1980) 1.

Berezin F.A. : Feynman path integral in a phase space. *Sov. Phys. Uspekhi* **23** (1981) 763.

Pervushin V.N., Reinhardt H. and Ebert D. : Path integrals in collective fields and applications to nuclear and hadron physics. *Sov. J. Part. Nucl.* **10** (1979) 444.

11.3. Symmetries

Rosen J. : Resource Letter SP-2: Symmetry and group theory in physics. *Am. J. Phys.* **49** (1981) 304. Erratum: **49** (1981) 793.

Michel L. : Symmetry defects and broken symmetry. Configurations. Hidden symmetry. *Rev. Mod. Phys.* **52** (1980) 617.
Fradkin E.S. and Palchik M.Ya. : Recent developments in conformal invariant quantum field theory. *Phys. Rep.* **44** (1978) 249.
Pignet O. and Ronet A. : Symmetries in perturbative quantum field theory. *Phys. Rep.* **76** (1981) 1.

Konopel'chenko B.T. and Rumer Yu.B. : Atoms and hadrons (classification problems). *Sov. Phys. Uspekhi* **22** (1982) 837.

Lopushan'ski Ya.T. : On symmetry in quantum field theory. *Fiz. Elem. Chastits At. Yadra* **13** (1982) 40 (in Russian).
Dao Vong Dyc: Conformal invariance in elementary particle physics. *Sov. J. Part. Nucl.* **9** (1978) 410.
Konopel'chenko B.G. : Symmetry groups in quantum field theory. *Sov. J. Part. Nucl.* **8** (1977) 57.

11.4. Gauge theories (see also 3.1, 4.1, 5.5, 7)

't Hooft G. : Gauge theories of the forces between elementary particles. *Sci. Am.* **242**, no.6 (1980) 90.

Gaillard M.K. : The unification of elementary forces. *CONAPP* **9** (1980) 39.

Jackiw R. : Introduction to the Yang–Mills quantum theory. *Rev. Mod. Phys.* **52** (1980) 661.

Narison S. : Technique of dimensional regulatization and the two-point functions of QCD and QED. *Phys. Rep.* **84** (1982) 263.

Berestetskii V.B. : Zero-charge and asymptotic freedom. *Sov Phys. Uspekhi* **19** (1976) 934.

11.5. Classical solutions of gauge equations

Actor A. : Classical solutions of SU(2) Yang–Mills theories. *Rev. Mod. Phys.* **51** (1979) 461.

Chao hao Gu: On classical Yang–Mills fields. *Phys. Rep.* **80** (1982) 251.

Leznov A.N. and Savel'ev M.V. : Exact solutions for cylindrically symmetric configurations of gauge fields. II. *Sov. J. Part. Nucl.* **12** (1981) 48.

Leznov A.N. and Savelyev M.V. : Exact cylindrically symmetric solutions to the classical equations of gauge theories for arbitrary compact Lie groups. *Sov. J. Part. Nucl.* **11** (1980) 14.

Leznov A.N. and Savel'ev M.V. : Some aspects of the theory of representations of semisimple Lie groups. *Sov. J. Part. Nucl.* **7** (1976) 22.

11.6. Summation of diagrams. Renormalization group

Maris H.J. and Kadanoff L.P. : Teaching the renormalization group. *Am. J. Phys.* **46** (1978) 652.

Kadanoff L.P. : The application of renormalization group techniques to quarks and strings. *Rev. Mod. Phys.* **49** (1977) 267.

Zinn-Justin J. : Perturbation series at large orders in quantum mechanics and field theories: application to the problem of resummation. *Phys. Rep.* **53** (1979) 157.

Vladimirov A.A. and Shirkov D.V. : The renormalization group and ultraviolet asymptotics. *Sov. Phys. Uspekhi* **22** (1979) 860.

11.7. Topology and differential geometry in physics

Bernstein H.J. and Phillips A.V. : Fiber bundles and quantum theory. *Sci. Am.* **245**, no.1 (1981) 94.

Parsa T. : Topological solutions in physics. *Am. J. Phys.* **47** (1979) 56.

Singer M. : Differential geometry, fiber bundles and physical theories. *(Phys. Today* **35** no. 3 (1982) 41.

Daniel M. and Viallet C.M. : The geometrical setting of gauge theories of the Yang–Mills type. *Rev. Mod. Phys.* **52** (1980) 175.

Mermin N.D. : The topological theory of defects in ordered media. *Rev. Mod. Phys.* **51** (1979) 591.

Madore J. : Geometric methods in classical field theory. *Phys. Rep.* **75** (1981) 125.

Eguchi T. , Gilkey P.E. and Hanson A.J. : Gravitation, gauge theories and differential geometry. *Phys. Rep.* **66** (1980) 213.

Ol'shanetskii M.A. : A short guide to modern geometry for physicists. *Sov. Phys. Uspekhi* **25** (1982) 123.

Perelomov A.M. : Solutions of the instanton type in chiral models. *Sov. Phys. Uspekhi* **24** (1981) 645.

11.8. Nonlinear equations and solitons

Jackiw R. and Rebbi C. : Topological solutions and instantons. *CONAPP* **8** (1978) 129.

Jackiw R. : Quantum meaning of classical field theory. *Rev. Mod. Phys.* **49** (1977) 681.

Olshanetsky M.A. and Perelomov A.M. : Classical integrable finite-dimensional systems related to Lie algebras. *Phys. Rep.* **71** (1981) 313.

Makhankov V.G. : Dynamics of classical solitons (in non-integrable systems). *Phys. Rep.* **35** (1978) 1.

Faddeev L.D. and Korepin V.E. : Quantum theory of solitons. *Phys. Rep.* **42** (1978) 1.

Izergin A.G. and Korepin V.E. : A quantum method of inverse problems. *Fiz. Elem. Chastits At. Yadra* **13** (1982) 501.

Mel'nikov V.K. : The inverse scattering method in the theory of nonlinear evolution equations. *Sov. J. Part. Nucl.* **11** (1980) 487.

Filippov A.T. : Nontrivial solutions of nonlinear problems in field theory. *Sov. J. Part. Nucl.* **11** (1980) 273.

Wiesner J. et al. : Iterative methods for solving the inverse problems in potential scattering theory. *Sov. J. Part. Nucl.* **9** (1978) 291.

Barbashov B.M. and Nesterenko V.V. : Dynamics of a relativistic string. *Sov. J. Part. Nucl.* **9** (1978) 391.

11.9. Problems common for quantum field theory and statistical physics

Hallock H. : Resource Letter SH-I: Superfluid helium. *Am. J. Phys.* **50** (1982) 202.

Witten E. : Quarks, atoms and 1/N expansion. *Phys. Today* **33**, no.7 (1980) 38.

Thacker H.B. : Exact integrability in quantum field theory and statistical systems. *Rev. Mod. Phys.* **53** (1981) 253.
Dolgov O.V. , Kirzhnits D.A. and Maximov E.G. : On an admissible sign of the state dielectric function of matter. *Rev. Mod. Phys.* **53** (1981) 81.
Savit R. : Duality in field theory and statistical systems. *Rev. Mod. Phys.* **52** (1980) 453.
Wehrl A. : General properties of entropy. *Rev. Mod. Phys.* **50** (1978) 221.

Brezin E. , Gervais J.L. and Toulouse G. : Common trends in particle and condensed matter physics. *Phys. Rep.* **67** (1980) 1.
Barber M.N. : Phase transitions in two dimensions. *Phys. Rep.* **59** (1979) 375.
Brezin E. and Gervais J.L. : Non-perturbative aspects in quantum field theory. *Phys. Rep.* **49** (1978) 91.

Anisimov M.A. , Gorodetskii E.E. and Zaprudskii M.M. : Phase transitions with coupled order parameters. *Sov. Phys. Uspekhi* **24** (1981) 57.
Patashinskii A.Z. and Pokrovskii V.L. : The renormalization-group method in the theory of phase transitions. *Sov. Phys. Uspekhi* **20** (1977) 31.

Kochetov E.A., Kuleshov S.P. and Smondyrev M.A. : Functional variational approach to polaron-type models. *Fiz. Elem. Chastits At. Yadra* **13** (1982) 635.

11.10. Computers and field theory. Gauge fields on lattices (see also 4.1)

Wilson K.G. : Problems in physics with many scales of length. *Sci. Am.* **24**, no.2 (1979) 140.

Creutz M. : Roulette wheels and quark confinement. *CONAPP* **10** (1981) 163.

Kogut J.B. : An introduction to lattice gauge theory and spin systems. *Rev. Mod. Phys.* **51** (1979) 659.

Creutz M. : Feynman rules for lattice gauge theory. *Rev. Mod. Phys.* **50** (1978) 561.

Drouffe J.M. and Itzykson C. : Lattice gauge fields. *Phys. Rep.* **38** (1978) 133.

11.11. Stochastic behavior of dynamic systems

Hofstadter D.R. : Metamagical themas. Strange attractors: mathematical patterns delicately poised between order and chaos. *Sci. Am.* **245**, no.5 (1981) 16.

Zeeman E.C. : Catastrophe theory. *Sci. Am.* **234**, no.4 (1976) 65.

Swinney H.L. and Golub J.P. : The transition to turbulence. *Phys. Today* **31**, no.9 (1978) 41.

Landauer R. : Stability in the dissipative steady state. *Phys. Today* **31**, no.11 (1978) 23.

Ott E. : Strange attractors and chaotic motions of dynamical systems. *Rev. Mod. Phys.* **53** (1981) 655.

Eckmann J.P. : Roads to turbulence in dissipative dynamical systems. *Rev. Mod. Phys.* **53** (1981) 643.

Zaslavsky G.M. : Stochasticity in quantum systems. *Phys. Rep.* **80** (1982) 157.

de Witt-Morette C. and Elworthy K.D. : New stochastic methods in physics. **77** (1981) 121.

Gaponov-Grekhov A.V. and Rabinovich M.I. : L.I. Mandel'shtam and the modern theory of nonlinear oscillations and waves. *Sov. Phys. Uspekhi* **22** (1979) 590.

Namsrai Kh. : Stochastic mechanics. *Sov. J. Part. Nucl.* **12** (1981) 449.

11.12. Axiomatic quantum theory

Ivanov S.S., Petrina D.Ya. and Rebenko A.L. : S-matrix in constructive quantum field theory. *Sov. J. Part. Nucl.* **7** (1976) 254.

Todorov T.S. : The algebraic approach in axiomatic relativistic quantum theory. *Sov. J. Part. Nucl.* **7** (1976) 94.

12. History and Prospects

Gingerick O. : The Galileo affair. *Sci. Am.* **247**, no.2 (1982) 119.
Kelves D.J. : Robert A. Millikan. *Sci. Am.* **240**, no.1 (1979) 118.

Hanle P.A. : The Schrödinger–Einstein correspondence and the sources of wave mechanics. *Am. J. Phys.* **47** (1979) 644.
Chandrasekhar S. : Einstein and general relativity. Historical perspectives. *Am. J. Phys.* **47** (1979) 212.
Wilson R. : From the Compton effect to quarks and asymptotic freedom. *Am. J. Phys.* **45** (1977) 1139.
Gaisser J.H. and Gaisser T.K. : Partons in antiquity. *Am. J. Phys.* **45** (1977) 439.

Brown L.M. and Hoddeson L. : The birth of elementary particle physics. *Phys. Today* **35**, no.4 (1982) 36.
Weisskopf V.F. : The development of field theory in the last 50 years. *Phys. Today* **34**, no. 11 (1981) 69.
Ramsey N.F. : Physics in 1981 ± 50. *Phys. Today* **34**, no.11 (1981) 26.
Rossie B. : Early days in cosmic rays. *Phys. Today* **34**, no. 10 (1981) 35.
Yang C.N. : Einstein's impact on theoretical physics. *Phys. Today* **33**, no.6 (1980) 42.
Brush S.G. : Poincaré and cosmic evolution. *Phys. Today* **33**, no.3 (1980) 42.
Sachs R.G. : Structure of matter: a five year outlook. *Phys. Today* **32**, no.12 (1979) 25.
Chandreasekhar S. : Beauty and quest for beauty in science. *Phys. Today* **32**, no.7 (1979) 25.
White D.H. and Sullivan D. : Social currents in weak interactions. *Phys. Today* **32**, no.4 (1979) 40.
Hoffmann B. : Einstein the catalyst. *Phys. Today* **32**, no.3 (1979) 36.
Brown L.M. : The idea of the neutrino. *Phys. Today* **31**, no. 9 (1978) 23.
Block F. : Heisenberg and the early days of quantum mechanics. *Phys. Today* **29**, no.12 (1976) 23.

Protopopescu S.D. : International Conference on High Energy Physics at Lisbon. *CONAPP* **10** (1982) 289.
Pondrom L.G. : Twentieth International Conference on High-Energy Physics. *CONAPP* **9** (1981) 199.
Weiskopf V.F. : Personal impressions of recent trends in particle physics. *CONAPP* **9** (1980) 49.

Lee B.W. and Quigg C. : An experimental fable. *CONAPP* **6** (1976) 93.

Pais A. : Einstein and the quantum gravity. *Rev. Mod. Phys.* **51** (1979) 861.

van Hove L. and Jacob M. : Highlights of 25 years of physics at CERN. *Phys. Rep.* **62** (1980) 1.

Ginzburg V.L. : What problems of physics and astrophysics are of special importance and interest at present? (Ten years later). *Sov. Phys. Uspekhi* **24** (1981) 585.

Okun L.B. : Contemporary status and prospects of high-energy physics. *Sov. Phys. Uspekhi* **24** (1981) 341.

Kisch D. : JINR experiments at the IHEP accelerator (in progress and planned). *Sov. J. Part. Nucl.* **10** (1979) 214.

SUBJECT INDEX

Abelian symmetry 14, 152, 176
accelerators, energy limits 100
action 5, 6, 157
angle
— θ 52, 132
— θ_c 47
— θ_w 65, 86
angles θ_1, θ_2, θ_3 48
annihilation 127, 128, 141, 142
anomaly 128
— of the axial current 129
— of the energy–momentum tensor 129
anticommutator 95
antiparticle 130
antisymmetric tensor
— —, three-dimensional 23, 88
— —, four-dimensional 52
asymptotic freedom 37, 38, 131
axial current 49, 128, 132
axial vector 49, 132
axion 53, 132
α_s 37, 52, 82, 116, 129, 131
α_w 82, 116
α_{em} 82
α_{GU} 82

Bag model 42, 133
baryon
— number 91, 130, 133, 164
— asymmetry of the universe 101
baryonium 42, 133
baryons 133, 153
Big Bang 12, 33, 99–109
b hadrons 31, 32, 133

Bjorken scaling 133, 144
blackbody radiation 134
black hole 134
Bohm-Aharonov effect 145
bosons 8, 135
bottomonium 32, 133
b quark 31, 133
branching ratio 143
bubbles 104, 108
β-decay 43
—, double 58, 145

Cabibbo angle 47, 48
Centauros 140
CERN 69, 135
charge conjugation 51, 58, 165, 135
— —parity 51, 52, 165, 135
charged current 44–50, 66, 136
charm 30
charmed particles 31
— quark 30
charmonium 31, 137
chiral symmetry (chiral invariance) 40, 128, 137
— super multiplet 175
classification 138
collaboration 113, 138
colliders 13, 69, 74, 100, 113
color 34, 35
— symmetry 36
commutator 158
condensate 42, 72, 74, 89, 102, 150, 169
conjugate currents 44, 50
confinement 20, 38, 40, 130, 138

conformal symmetry 97, 139
conservation laws 6, 4, 157
coronas of galaxies and of clusters of galaxies 56, 100
Coleman-Weinberg potential 76, 108
combined parity 165
continual integral 149
cosmic rays 139
cosmological term 97, 106
cosmology 99–109
CP symmetry 51, 52
— violation 52, 75, 146, 165
CPT symmetry 51
— theorem 51, 96, 140
cross section 140
— — differential 141
— —, elastic 141
— —, exclusive 141
— —, inclusive 141
— —, semi-inclusive 141
— —, total 141
cryptoexotic baryons 42, 147
— mesons 42, 147
current 142
—, negative 44
—, positive 44

Decay rate 142
decuplet 25, 29
deep inelastic processes 13, 38, 143
de Sitter universe 108
DESY 144, 166
diagonal 68, 144
dimension 116, 157
dimensional estimates
— — in cosmology 100, 106
— — — mass hierarcy problem 94
— — for black holes 134
— — — coupling constants 117
— — — cross sections 140–141
— — — cosmological term 97, 106
— — — monopole mass 88–90
— — — neutron oscillations 92, 165
— — — photon gas density 106, 134
— — — proton lifetime 85
— — — weak decays 44
Dirac mass 58
— matrices 14, 50

— string 89
double β-decay 58, 145
dyon 146
Δ-baryons 20

E(6) symmetry 92
E(7) symmetry 93
E(8) symmetry 93
Einstein-Podolsky-Rosen paradox 146
electric dipole moment of a particle 52, 75, 146
electromagnetic interaction 10, 12, 14
electron volt 117
electroweak symmetry 64
— theory 63–79
Eötvös experiment 111
equations of motion 6
Euclidean space 4
exceptional groups 92, 160
exclusive cross section 141
exotic baryons 42, 147
— mesons 42, 147
extended supersymmetry 97, 98

Fermi constant, G_F 44, 50, 54, 67, 72, 78, 94, 117
fermions 8, 33, 44, 51, 64, 70, 91, 147
Feynman diagrams 8, 14, 148
— scaling 148
FIAN 149
field 5, 11, 36, 65, 70–77, 83, 88, 168
—, physical 5, 168
flavor 32, 68, 149
FNAL 69, 149
four-fermion interaction 44, 50, 54, 67, 78
Friedmann universe 12, 107
Froissart limit 141
functional 149
fundamental interactions 10
— Lagrangian 6, 77, 157
—particles 19
— representation 160

Gauge desert 93
— fields 36, 65, 70, 76, 88, 97
— super multiplet 175
— symmetry 14, 64, 89, 97, 149, 159
Gell-Mann-Low function 129, 131

SUBJECT INDEX

Gell-Man matrices 28, 84
general relativity 11, 78, 110, 151
generations of fermions 33, 47
global symmetry 159
glueball 42, 150
gluodynamics 129, 150
gluino 98
gluon condensate 42, 150
gluons 34–42, 151
Goldstone bosons 41, 70
grand unification 82–109
gravitation 10, 110, 151
gravitational constant 11, 94, 117, 151
gravitino 96, 98
graviton 12, 96, 98, 152
gravity waves 12, 152
groups 152, 159

Hadron 10, 19–41, 153, 162
— jets (quark and gluon jets) 38
"hairpin" 21
hard processes 144
hedgehog 88, 89
helicity 40
— multiplets 64, 83
— states 50, 64, 143
Heisenberg uncertainty relation 112
Hermitian conjugation 22, 160
Hilbert space 7, 169
Higgs bosons 74, 153
— mechanism 70, 174
horizontal bosons 73
—symmetry 93
Hubble constant 106
hypercharge 26, 64, 153
hyperon 25, 28, 130, 153

Identical elementary particles 153
IHEP 154
inclusive cross section 141
inflationary universe 108
"ino" 98
instanton 42, 154
intermediate bosons 63, 79, 113, 156
internal symmetries 7, 95, 175
invariance 4, 14, 37, 173
invariant charge 131
invisible mass in the universe 100, 106, 107

isotopic space 163
— spin in strong interaction 22, 163
— — — electroweak interaction 54, 64
— symmetry 22, 163
ITEP 55, 156
ITP 156
I_3Y diagrams 26–29

JINR 156

K mesons 24, 48, 51, 75
KNO scaling 156
Kobayashi–Maskawa matrix 48, 136

Lagrange function 5, 157
Lagrangian 6, 13, 14, 36, 50, 54, 157
Lattices 139
least-action principle 5, 6, 157
left-handed particles 40, 50, 54, 64, 68, 75, 83, 84
LEP 70, 74, 156, 158
leptonic number 58, 59, 92, 158
leptons 10, 32, 34, 158
Lie algebras 158
— groups 159
lifetime 143
local symmetry 23, 56, 79, 159
Lorentz group 173, 174
— transformation 173
low-background laboratories (neutrino observatory) 86, 87, 172
lower leptons 33, 54
— quarks 33, 54
Lüders-Pauli theorem (CPT theorem) 51, 96, 140
luminosity 160, 161
Λ_{QCD} 38, 130, 131

Magnetic monopole 88, 145
Majorana mass 58, 60, 161
— neutrino 58, 59, 161
majoron 161
mass 5, 72, 74, 161, 162
— of the neutrinos 54, 56, 161
meson 19–41, 153, 162
— factory 93
microwave background radiation 100

Minkowski space 4, 173
mirror asymmetry 48, 64, 75
— reflection of coordinate axes 49, 51
— symmetry 64, 75
mixing 28
multiplet 23, 25, 162
— of hadrons 162, 163
muon 44, 66, 77, 93, 162
"mu-meson" 162

Neutral currents 53, 68, 142, 164
neutrino 10, 44, 53–60, 164
— oscillations 56, 164, 172
— reactions 44, 53
neutron 20
— oscillations 92, 164
Newton's constant 11, 12, 117, 151, 152
non-Abelian gauge fields 36, 42, 84, 85
nonperturbative effects 145, 163, 167
nucleons 9, 20, 153, 162

Occam's razor 110, 164
octet 25, 29, 151
operators of creation and annihilation of particles 7, 13, 44, 45
oscillations of K mesons 56, 164
— — neutrinos 56, 58, 164, 172
— — neutrons 92, 164

Parity 48, 165
— violating nuclear forces 48, 166
— violation 48, 51, 165, 166
Pauli matrices 23, 84, 158
PEP 166
PETRA 53, 144, 166
phenomenology 166
photino 98, 99
photon 166
physical vacuum 48, 72, 89, 103, 108, 176
pions (π mesons) 21, 40, 45, 46
Planck mass 167
Planck's constant 7, 117, 167
Poincaré group 6, 159, 173
polarization of vacuum 17, 82
Pomeranchuk theorem 167
positron 13, 15, 17
positronium 136, 167
powers of ten 119

P parity 48, 165
P reflection 49, 51, 165
primordial radiation (microwave background) 100
probability amplitude 167
propagator 15, 67, 148
proton–antiproton collider 65, 74, 113
proton decay 85, 90

Quantum 168
— chromodynamics (QCD) 36, 168
— electrodynamics (QED) 12, 14, 41, 52, 168
— field theory 168
— mechanics 6, 8, 168
— of action 7, 168
quark condensate 169
— diagrams 21, 25
quarks 14–43, 170
quasistable particles 9, 153

Resonances 9, 20, 153
right-handed particles 40, 50, 54, 68, 75, 83, 84
rotated quarks 47, 64, 136
running constants 17, 82, 99
ρ mesons 26, 27

Scalar bosons (Higgs bosons) 74, 75, 153
scaling, Bjorken 133
—, Feynman 148
—, KNO 156
SI (International System of Units) 118
singlet 25, 167, 171
SO(10) symmetry 91, 92
solar neutrinos 171
special relativity 4, 172
spin 8, 111
spinor 22
spontaneous symmetry breaking 41, 70, 72, 174, 175
stable particles 9
state vector 7, 169
sterile neutrino 164
strangeness 24
strange particles 24, 25, 47
strings in QCD 39
strong interaction 9, 19–42

SU(2) symmetry 22, 23
SU(3) symmetry 25, 30, 35
SU(5) symmetry 83–93
supergravity 96
supermultiplet 95, 175
superparticles 98, 99
supersymmetry 95–99
superunification 98
symmetries, role of 3, 79
system of units $\hbar = c = 1$ 115, 116
systems of units 115–125

Technicolor 76
technigluons 76
technipions 76
techniquarks 76
time reversal 51, 140, 146
toponium 32
$\theta_1, \theta_2, \theta_3$ angles 48
θ_c angle 47
θ_w angle 65, 86
θ-term 52, 132
θ-τ paradox 48, 77
τ lepton 44

ultraviolet divergence 78
uncertainty relation 112
unimodulus matrix 22
unitarity 22, 160
unitary matrix 22, 160
— symmetry 22
— transformation 22, 160
universe
—, closed 105, 107
—, flat 105

—, hot 100
—, open 105, 107
upper lepton 33, 54
— quark 33, 54
upsilon 31, 32
upsilonium (the same as bottomonium) 32, 133

V–A current 49, 50
vacuum angle 52, 53
— bubbles 104, 108
— condensate 72, 102
— domain walls 104
— domains 103, 104
— expectation value 72, 74, 102, 104
— strings 103
vector current 49
— particles 26, 27
VEPP 176
virtual particle 15, 17, 176

Weak decays 43, 44
— interaction 43–79
— reactions 45, 46
Weinberg angle 54, 65–68, 86
Weyl neutrino 51

X bosons 84, 85

Yang-Mills fields 42, 176
Y bosons 84, 85
Yukawa coupling 71, 73, 176, 177
— — constants 71, 73, 74, 176, 177

Z boson 63–75, 79, 113, 156

Particle Physics
The Quest for the Substance of Substance

by L.B. Okun, Institute of Theoretical and
Experimental Physics, Moscow
Translated from the Russian by V.I. Kisin

Volume 2 in the Contemporary Concepts in Physics series

Written by one of the world's leading theoretical physicists, this comprehensive volume offers a thorough overview of elementary particle physics and discusses progress in the field over the past two decades. **Particle Physics** forges links between new theoretical concepts and long-established facts in a style that both experts and students will find readable, informative, and challenging. A special section explains the use of relativistic quantum units, enabling readers to carry out back-of-the-envelope dimensional estimates. This ambitious book opens the door to a host of intriguing possibilities in the field of high-energy physics.

Other Harwood books of interest

PARTICLE PHYSICS AND INTRODUCTION TO FIELD THEORY
by T.D. Lee

HIGH ENERGY PHYSICS WITH NUCLEI
by Yu.P. Nikitin and I.L. Rozental

QUANTUM MECHANICS AND NONLINEAR WAVES
by Philip B. Burt

ELECTRON AND PION INTERACTIONS WITH NUCLEI AT INTERMEDIATE ENERGIES
by William Bertozzi, Sergio Costa and Carlo Shaerf

ISSN: 0272-2488
ISBN: 3-7186-0228-8 (CLOTH)
3-7186-0229-6 (PAPER)

G+B/harwood
chur • london • paris • new york